On the
Surface
of Things

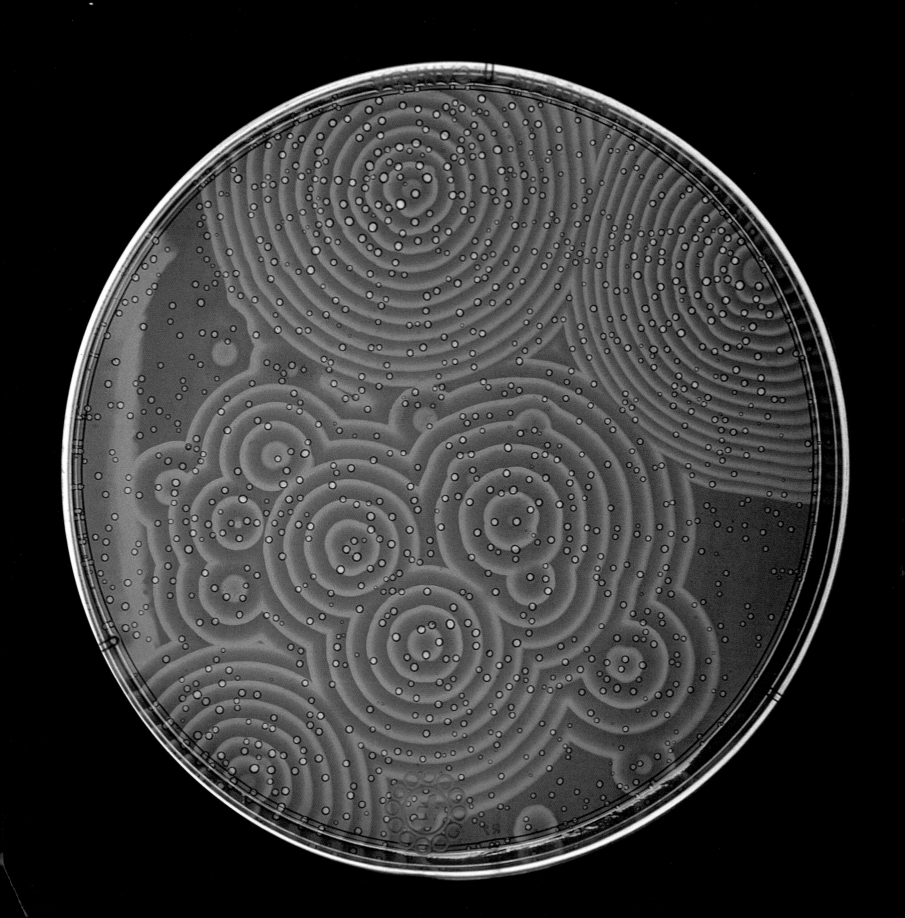

FELICE FRANKEL AND GEORGE M. WHITESIDES

On the Surface of Things

IMAGES OF THE EXTRAORDINARY IN SCIENCE

HARVARD UNIVERSITY PRESS
Cambridge, Massachusetts
London, England
2007

First published in the United States by Chronicle Books.

First Harvard University Press paperback edition, 2007

Image facing title page: *Patterns from an Oscillating Chemical Reaction*
Image at right: *Etched Silicon*
Image facing introduction: *Polymer-containing Gel*

Book design: Stuart McKee

Library of Congress Cataloging-in-Publication Data
Frankel, Felice.
    On the surface of things: images of the extraordinary in science
 / Felice Frankel and George M. Whitesides
        p.   cm.
    Includes bibliographical references and index.
    ISBN-13: 978-0-674-02688-9 (pbk.)
    ISBN-10: 0-674-02688-8 (pbk.)
    1. Surfaces (Physics) 2. Optical Images. 3. Surfaces
(Physics)—Pictorial works. I. Whitesides, G. M. II. Title
    QC173.4.S94F73 1997
    530.4' 17—dc21   97–852

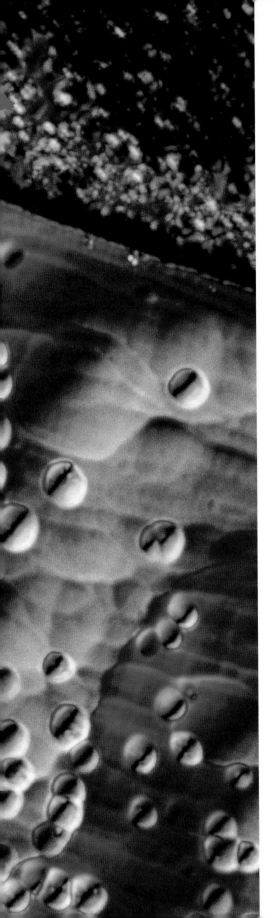

contents

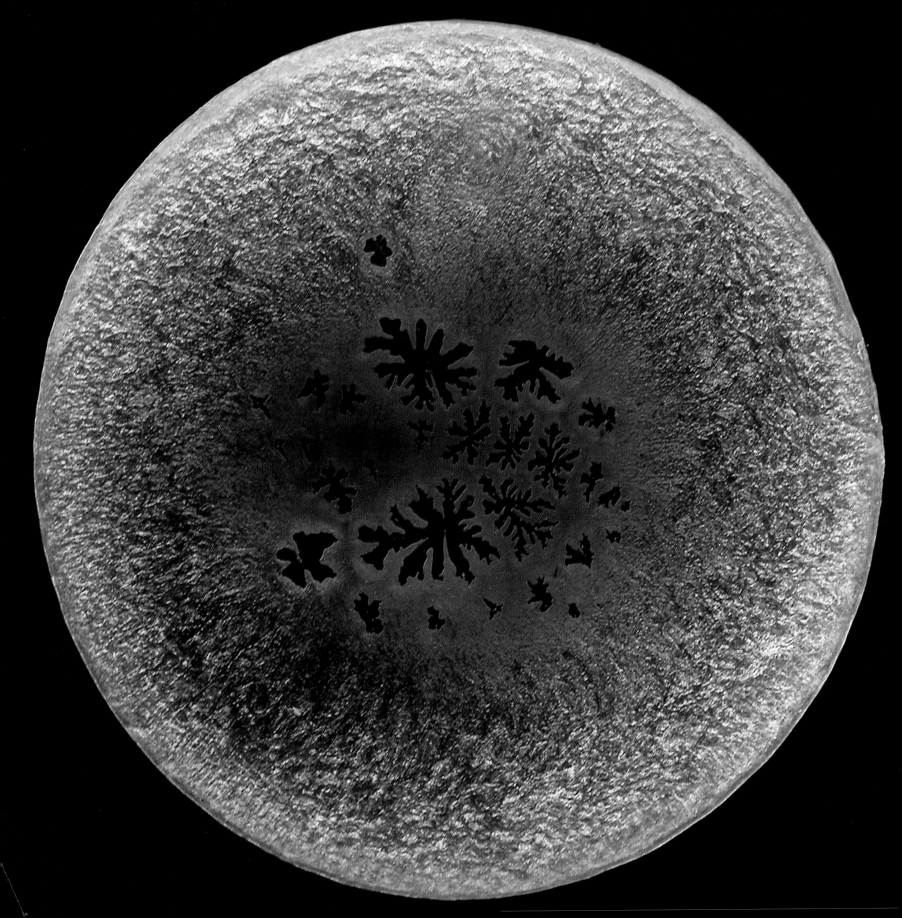

introduction

Surfaces define the shapes of our world; light allows us to see them. This book is about surfaces, and light, and how they interact.

Understanding the physical world—science—is a joy. Deciphering its rules provides puzzles for a lifetime; these rules also nourish the technologies that enrich and vex us. Nature has the capability to beguile endlessly, and in describing nature, science shares that capability. But science is also a rose bush: the flowers are ravishing, but protected by thorny technical details. In this book, we try to display some of the flowers and avoid most of the thorns.

The book originated in 1997 in an improbable coincidence of interests in science and images. One of us, Felice Frankel, is a photographer who has worked with the interaction of nature and humanity on a grand scale—in landscapes. She became enthralled by the possibilities of powerful imaging in science, and by the general ineptitude of scientists in exploiting those possibilities. The other, George Whitesides, is a chemist who works at the other end of the scale of sizes, making molecules and microstructures.

In his years of teaching university students Whitesides learned, among other lessons, that equations and scientific graphics are far from riveting. We agreed that arresting images were uniquely effective in describing nature. A single image organizes a deluge of information in a form that is easy to understand. An image that is rich in composition and color always catches the eye. And what catches the eye, catches the mind.

The photographer's optimistic view was that it is possible to make any subject visually arresting; the scientist's opinion was that it might be harder than it seemed. We agreed to try the experiment, and this book is the result. This tenth anniversary edition has several new images but retains the same spirit. We chose the subject—surfaces, and the light that illuminates them—because surfaces are what we see of objects, and because they are important in the two great technologies—information and biology—that are now remaking the world.

Information technology promises that computers will let us find what we want: one another, wealth, conflict, opportunity, dominance, equality. Biology views us, and all living things, as wondrously complicated machines: collections of devices

7

patched together by evolution, with myriads of molecular parts and with instruction books—the genomes—open to view. As a society, we are of mixed mind about both. Access is what we gain from them—to one another, to knowledge, to opportunity, to health. Privacy—the separateness of self—and the mystery of "life," are what we lose. Our concept of a person is eroding in a storm of bits.

Both computers and living organisms are built of small things, and surfaces are especially important for them. A transistor—the building block of computers—is a tissue of invisible surfaces spun into silicon. A living cell—the universal unit of life—is a collection of molecules that interact through the touch of their surfaces. Surfaces are also, in some sense, a separate state of matter, distinct from solids, liquids, and gases: surfaces are where the properties of matter *change* most rapidly.

And as for light: We are a species that specializes in seeing. Sight is our clearest window on the world; it brings us our terrors and our delights. We would not forego seeing the faces of our children, or the rose of dawn, or the leaves spinning in the autumn wind, but the same eyes bring us the shadows of our nightmares.

Beautiful images connect humanists and scientists. Both need a clearer understanding of the forces that shape our world. We offer these images and descriptions to anyone who wishes to know more about the tangible world. Technical and photographic notes and literature references end the book.

We invite you, the reader, to share our love of science, and images, and words.

the sizes of things

Without a sense of size, images are ambiguous. A mammalian cell up close and a cluster of stars far away are both too small to see; magnified, but with no further indication of size, they may be difficult to tell apart. Our emotional response to a polar bear and to a shrew—both ferocious predators—is based on knowing our size relative to theirs.

The images in this book are difficult to interpret without knowing their sizes. We use the metric system to describe them for two reasons. First, the metric scale is universally used in science and technology. Second, it has the convenient feature that all lengths are given as fractions or multiples of 10 and are based on a single unit of a single length—the meter.

Most of the images in this book are of objects smaller than a meter. A meter is approximately a yard—about three feet. A centimeter is one hundredth of a meter, or approximately one-half inch. A millimeter is one thousandth of a meter—the size of a pinhead. A micrometer (often called a micron) is one millionth of a meter, or one thousandth of a millimeter. A one-micrometer object is too small to see without magnification: the period at the end of a sentence is approximately 500 micrometers in diameter; a fine hair is approximately

100 micrometers in diameter; a red blood cell, which is invisible to unaided sight, is 10 micrometers in diameter; visible light has wavelengths of about one-half of a micrometer. The smallest measure we use is a nanometer—one billionth of a meter. An atom is two-tenths of a nanometer in diameter.

The smallest features in the images in this book are about a nanometer in size; the largest are approximately 10 meters. Most of the objects are between 1 micrometer and 10 centimeters in size—a factor of 100,000. To avoid the mental arithmetic of estimating sizes for each object from words, we have supplied a familiar shape—the head of a pin—to use as a reference. (A pinhead is a circle one to two millimeters in diameter.) For images of large things, the relative size of the pinhead disappears to a dot; for images of very small things, the relative size of the pinhead may be much larger than the page, and we show only a section of its circumference. A quick comparison of the size of the object to the size of the pinhead should help readers imagine the proper scale for each image. Further details about some of the sizes are included in the technical descriptions in the "Notes and Readings" section at the back of the book.

○   DIAMETER OF A PINHEAD AT ACTUAL SIZE

for images of large things, the relative size
of the pinhead disappears to a point

DIAMETER OF A PINHEAD MAGNIFIED 10 TIMES

DIAMETER OF A PINHEAD MAGNIFIED 50 TIMES

11

DIAMETER OF A PINHEAD MAGNIFIED 125 TIMES

DIAMETER OF A PINHEAD MAGNIFIED 1000 TIMES

for images of very small things, the relative size
of the pinhead may be much larger than the page,
and we show only a section of its circumference

light

light

is the insubstantial
foundation of our world.

Light is the chirps made by electrons as they change: from one position to a more stable one, from one speed to another. These chirps are brief, shrill warbles of electromagnetic radiation. The chirping of electrons is profoundly different from the chirping of birds — quantum phenomena influence the interactions of light with matter in ways that have no perfect analogy in everyday experience. And yet they are also similar in that both involve waves — for light, waves of electromagnetic radiation; for sound, waves of pressure in air.

The terrain through which light travels is space and matter. To light, matter is pools of electrons collected around atomic nuclei. Where the pools are deep, light travels slowly; where the pools are shallow, it moves rapidly. Most phenomena involving light — bending, reflection, absorption — result from encounters of waves of light with the pools of electrons in matter. Light interacts strongly only with electrically charged particles, and especially with electrons: we perceive this interaction primarily as the absorption and emission of light in the colors of the visible spectrum and as heat.

Light is the insubstantial foundation of our world. The energy from the sun that fuels life arrives as light, both the colors we see and the forms of electromagnetic radiation that we usually do not think of as "light" — radio waves, heat, ultraviolet light, and X rays. We see with light: it is the air perception breathes. We communicate and measure with light. We use light to talk, write, send pictures, and move objects, and to machine and weld matter.

A car drips dirty oil; a drop falls onto water in the mud. There it spreads and forms a rainbow skin as beautiful as that of any salamander. The drop is pulled nearly flat by the water. The molecules of water exposed at its surface with air are uncomfortable. Molecules are most stable when in contact with other similar molecules on all sides. Water molecules exposed to air are partially "bare," and thus higher in energy and less stable than those in the interior of the puddle. These surface molecules are more comfortable when covered with a skin of oil. Detergents added to the oil to help it cover surfaces in the engine of the car also help it to cover the surface of water. With enough time, the oil will be pulled to a thickness of only a few hundred molecules. A small quantity of oil will cover an amazingly large area: a tablespoon full, as a single molecular layer, would cover a surface the size of several football fields.

oil slick

The film of oil creates two imperfect mirrors—surfaces that are flat, approximately parallel, but only partly reflecting. A light wave first encounters the surface between the air and the oil; some of it reflects, some continues on into the oil and encounters the surface between the water and the oil. Here again, some reflects and some continues. When we look at the oil slick, we see the combination of the reflected light waves. Because the two mirrors are close—the distance between them is similar to the wavelength of light—light reflecting from them interferes with itself: that is, waves of light reflecting from one mirror augment or annihilate waves reflecting from the second.

The colors measure the thickness of the oil, with only one color reflecting for a particular thickness. Each passage from red to blue corresponds to the addition of approximately 200 molecules of oil to the thickness of the film. Amazingly, simple observation

can infer the structure of nature at the molecular scale: by looking at the colors, one can estimate changes in the thickness of an oil film from place to place to within perhaps 50 molecules, which is five hundred-thousandths of a millimeter (or two millionths of an inch).

Thin oily films are ubiquitous. These films contain molecules—hydrocarbons—that are chains of carbon atoms combined with hydrogen atoms. Such compounds are the basis of gasoline, lubricants, and soaps; they also occur naturally in the decay of plant matter in the absence of oxygen. Because they tend to form colorful films when they float on the surface of water, they are highly visible evidence of environmental contamination.

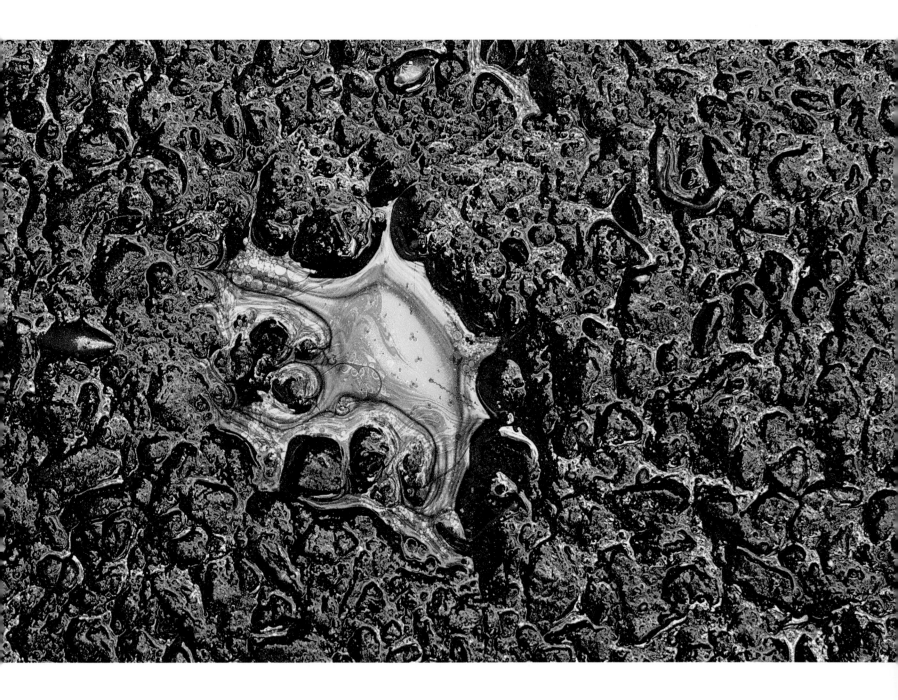

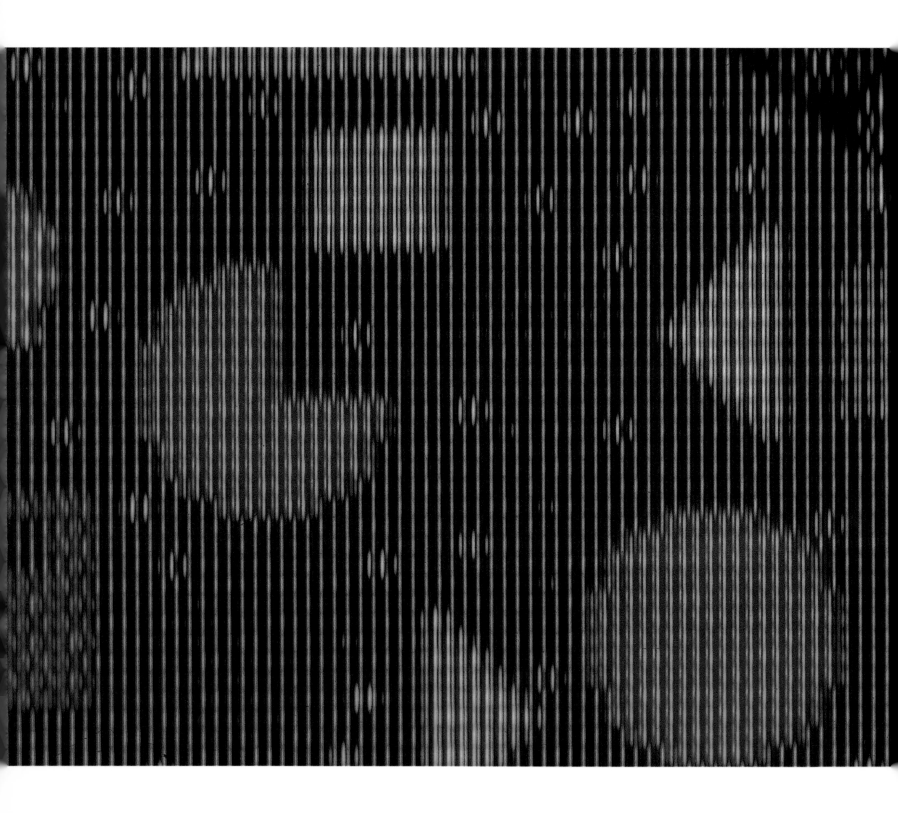

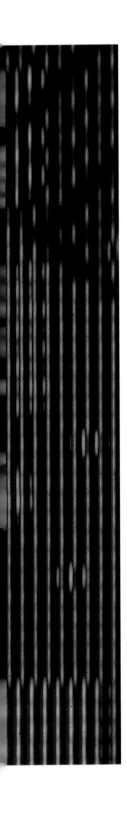

Light is hard to steer: it has no handles we can grasp—no mass, no charge. Electrons are more docile. They can be guided with electrical charge, toward positive charge and away from negative. A computer screen is a billboard on which we paint patterns of light using electrons as the brush. We guide the electrons, which we cannot see, to form the light, which we cannot guide.

A few monitors and television sets still use cathode ray tubes as their display screens. These are large vacuum tubes that generate a stream of electrons and spray it onto the light-generating surface. The stream is pumped from a wire into the vacuum and passes between charged metal plates. Moving the electrical charge between the plates steers the electrons. When the electrons careen into the screen, they strike light-generating atoms (phosphors). After impact, the electrons in the phosphors ring like the bells of a carillon: each kind vibrates in a characteristic color. (Plasma displays also use streams of electrons to generate light, although the details are different.) Only a few different phosphors are needed; their combination provides the full visible spectrum. Each glowing spot is the ghost of dying electrons, the transmogrification of kinetic energy into light.

Understanding the processes that cause the phosphors in the displays of computers and television sets to lose their colors, and the electron beam its accuracy, aids the design of brighter and longer-lived displays.

17

computer monitor screen

A pebble dropped in a still pond spreads circles of ripples. Matter is pools of electrons held in place by atomic nuclei. Electrons carry a negative electrical charge, and negative charges repel one another. When one electron moves, other electrons feel that movement. Light is the ripples when these pools of electrons are disturbed.

If you could glue an electron to a spring on the surface of the sun and set it to vibrating rapidly — 500,000 times per second — electrons at the surface of the earth would feel that vibration eight minutes later. What we call light is the influence of one vibrating electron on others. Here, it is light that vibrates too slowly for our eyes to recognize — we would call it radio waves — but it is light, nonetheless.

For a single vibrating electron, this light would be far, far too weak to be detected by any means. If it were made much stronger, by large numbers of electrons or ions (atoms missing electrons or attached to extra electrons) oscillating together, a radio could detect it. Only if the vibration were a billion times faster — 5×10^{14} or 500 million million times per second — would we perceive it as visible light.

The intricate interaction we call seeing is electrons vibrating in response to other electrons. When we see, electrons in our eyes — in molecules in the retina — vibrate in response to movements of electrons in the scene on which we gaze; this vibration is translated into messages to the brain. When we see the sun, we are sensing the violent motions of electrons and ions on its surface. When we see the reflection of the sun on water, we are seeing the movements of electrons in the water, responding to movements of electrons in the sun.

The surface of the sun is not, of course, smoothly vibrating electrons and ions: it is cacophony, a superheated stew of electrons and atomic nuclei at full boil, spewing light of all colors — radio waves, microwaves, heat, the visible colors, ultraviolet light, X and gamma rays — in all directions. At our safe distance, beneath the protective sea of our atmosphere, with our very limited eyesight, we see only the distant light that we mistake to be a pure color, "white."

18

reflections of
sunlight on water

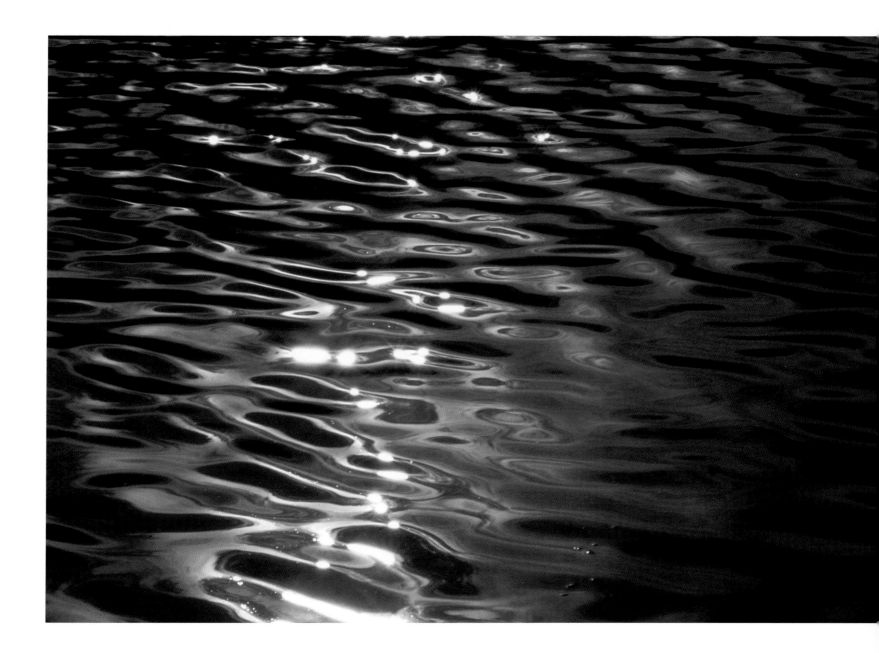

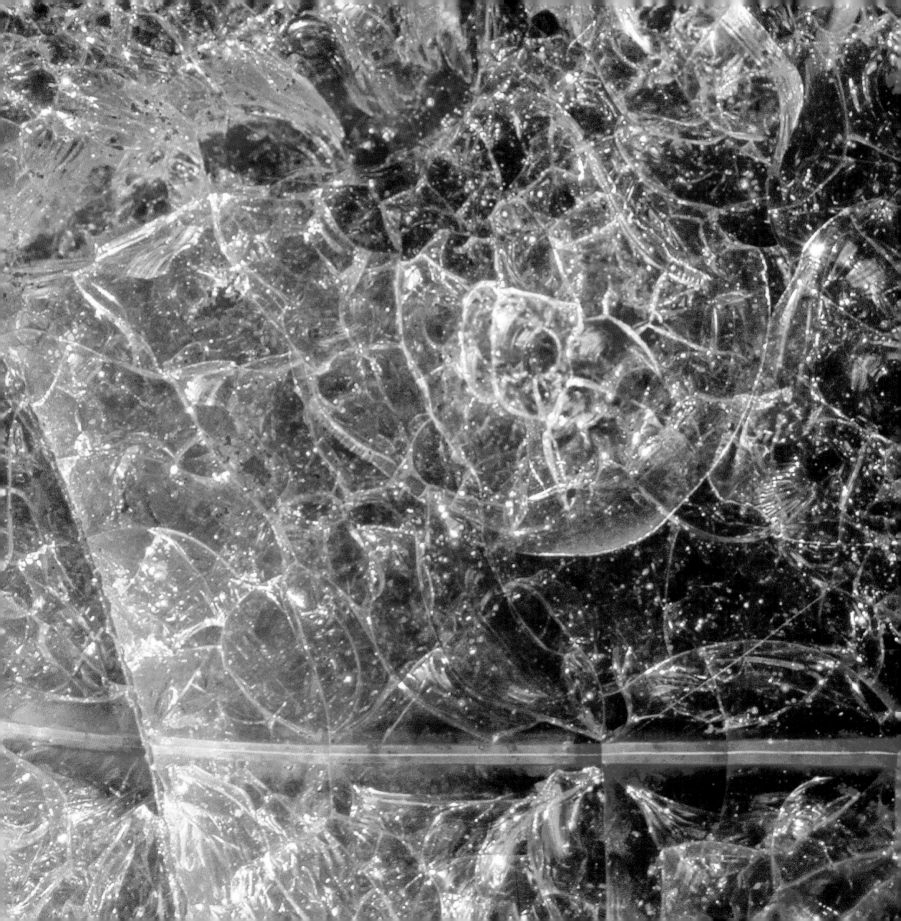

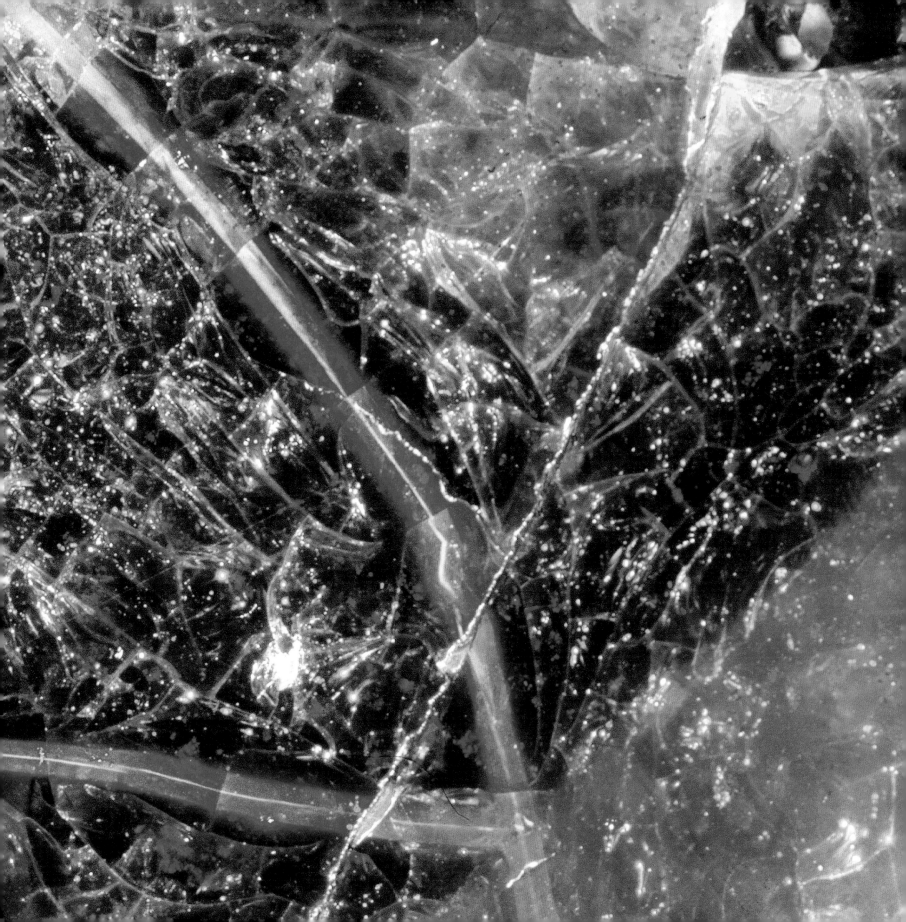

veins of opal

PREVIOUS PAGES

Opal is a complete surprise. It would be no less surprising to find necklaces of perfect pearls forming spontaneously in a drainage ditch. Opal is formed by simple heating and cooling of water in an underground crack, in a process that generates arrays of spheres—diffraction gratings—that separate white light into pure colors. From dilute mineral broth to rainbows frozen in rock: spontaneous generation of order from disorder.

Over time, cracks in rocks filled with hot water that slowly became saturated with silica extracted from the rock. Very, very slowly the water cooled. As it did, the silica precipitated as tiny, uniform, transparent spheres floating in the hot fluid like fish eggs. These spheres settled and arrayed themselves in beautifully ordered sheets spaced apart by a few hundred nanometers—that is, by 200 billionths of a meter, or one ten-thousandth of an inch—coincidentally, the size of waves of visible light. The spaces between

the spheres filled with silica with slightly different properties, and this new material glued the arrays in place. To light, these sheets of spheres are semitransparent mirrors, and light reflects from them in a way that allows only single colors to escape. The origin of the colors of opal, an oil slick, and the wings of the morpho butterfly is the same: diffraction, waves of light reflecting from arrays of parallel planes and augmenting or destroying one another.

Opal demonstrates that three-dimensional, ordered arrays of small particles with remarkable optical properties can form spontaneously and without our participation. This type of process is called self-assembly. *We are not involved*—the process proceeds and generates a complex structure by itself. Designing self-assembling processes to make complex, functional materials according to our specifications is an adventurous new strategy now being explored in electronics and optics in the hope that it will simplify the manufacturing of these materials.

22

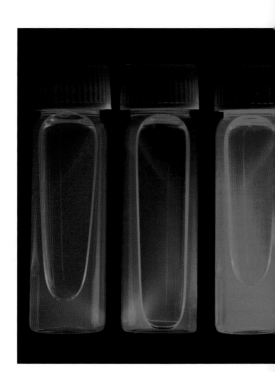

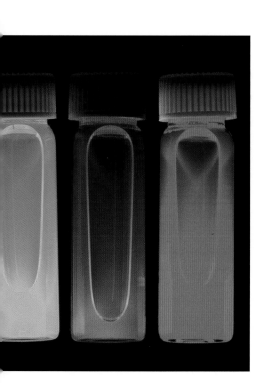

The pitch of organ pipes is determined by the length of the resonant tubes. The longer the pipe, the deeper the note. The color of the light given off by electrons in solid particles is also determined by the size of their box, the particle in which they vibrate. The larger the box, the lower the "pitch" and the slower the vibration. Ultraviolet light from electrons in a small box is too shrill—too fast—for our eyes to see. The vibrations slow through the colors blue, green, yellow, red, and finally basso purple, before the pitch falls into the infrared, invisible to the eye but perceptible to the skin as heat.

These vials contain suspensions of particles of cadmium selenide with diameters between 2 and 8 nanometers; the particles are "quantum dots" shaped to be carefully pitched organ pipes for electrons. The quantum dots are boxes just small enough to give electrons claustrophobia, each sized to give off a different color when ultraviolet light causes its electrons to vibrate. Ultraviolet light is energetic and causes the particles to give off bright visible light—fluorescence. With visible light, which is less energetic, the suspensions absorb rather than emit light, and they appear as pale, transparent pastels.

These colors may someday be used in computer display screens. The colors in current displays rely on the range of different phosphors in the screen. Generating a complete range of colors out of particles with different sizes of a single material may simplify the design and manufacture of these ubiquitous screens.

suspensions of small,
fluorescent particles

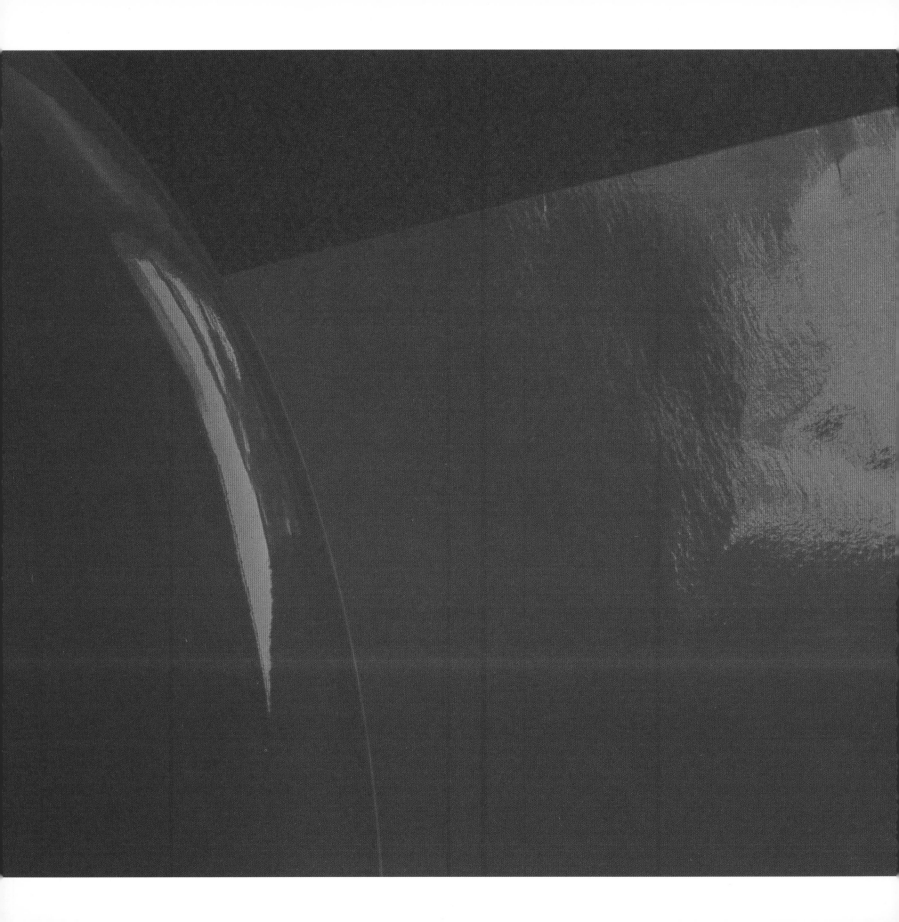

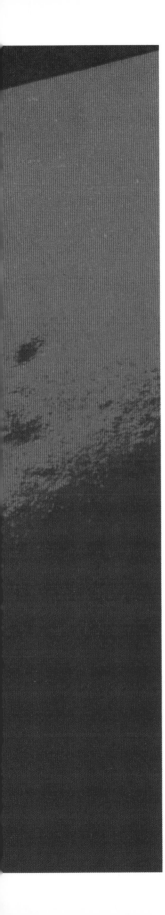

ence. The electrons in the molecules of a colored object respond to light of the same color by vibrating at the same frequency and emitting clearly the "tonal" wavelength that set them going. The electrons damp other colors — other frequencies — in the light; these colors are subtracted from the light emitted.

When we see a surface that has been illuminated with white light as red, the electronic tuning forks that oscillate at the frequency of red in the object are ringing clearly (albeit at 5×10^{14} ripples in the electromagnetic pool each second); the frequencies corresponding to yellow and blue are damped and their characteristic colors are sucked from the reflected light.

A madrigal is a rich polyphony. Silence all the parts but one and what remains is simple melody. "Color" is "white light" with most of the parts gone: the blessing of color comes by subtraction. A red rose is a sponge that sucks the blue and yellow from white light; this red plate and red paper trap all colors other than red. White light offers a palate of all colors to objects; they remove some, and their "color" is the light they reflect.

Seeing color has features in common with the sympathetic vibration of tuning forks. A tuning fork tuned to A (440 cycles per second) will ring sympathetically only to that tone. An electron in a molecule is similar: it can vibrate only at certain frequencies. Electrons in one molecule can respond to vibrations from a second by ringing in sympathy or by resting in indiffer-

the color red

tubular gels

Slap the face of a guitar sharply with your hand, and the strings will vibrate gently. The electrons in a molecule react the same way: slap them with ultraviolet light and they may answer with a gentler vibration—of visible light. Our eye cannot see the ultraviolet slap; we can only see the more slowly oscillating visible response.

These "tubes" are molded gels—gossamer, transparent networks of long, interconnected organic molecules swollen with water: a three-dimensional "fishing net" with strands a few tenths of a nanometer across. Molecules of fluorescent dyes are attached to these strands like fish caught in a net. In visible light, the networks of tubes are transparent, invisible to the human eye. Irradiating the dyes with ultraviolet light renders the translucent gels visible.

The gels are potentially useful to us because they can swell or shrink enormously—by as much as 1000 times their original volume. Small changes in the acidity of the solution permeating the gels bring about the great change in volume, and they can also change volume if salts—including table salt—are added to the water. The gels may be developed for trapping and releasing drugs, for soaking up water, or for serving as a kind of artificial "muscle" to move small objects in water.

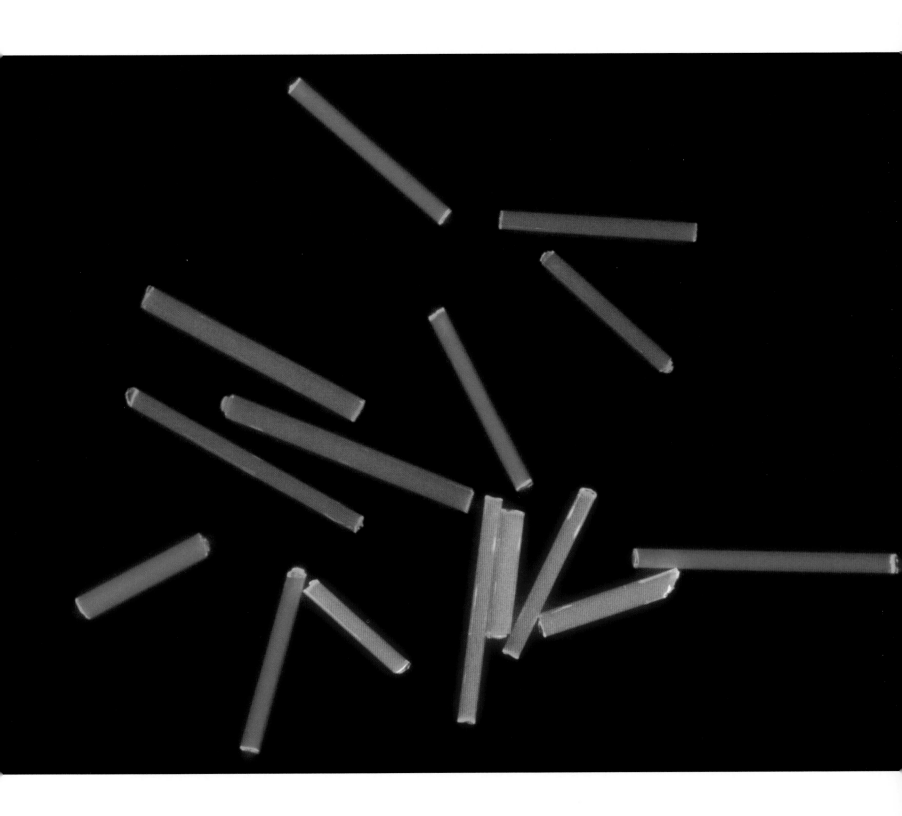

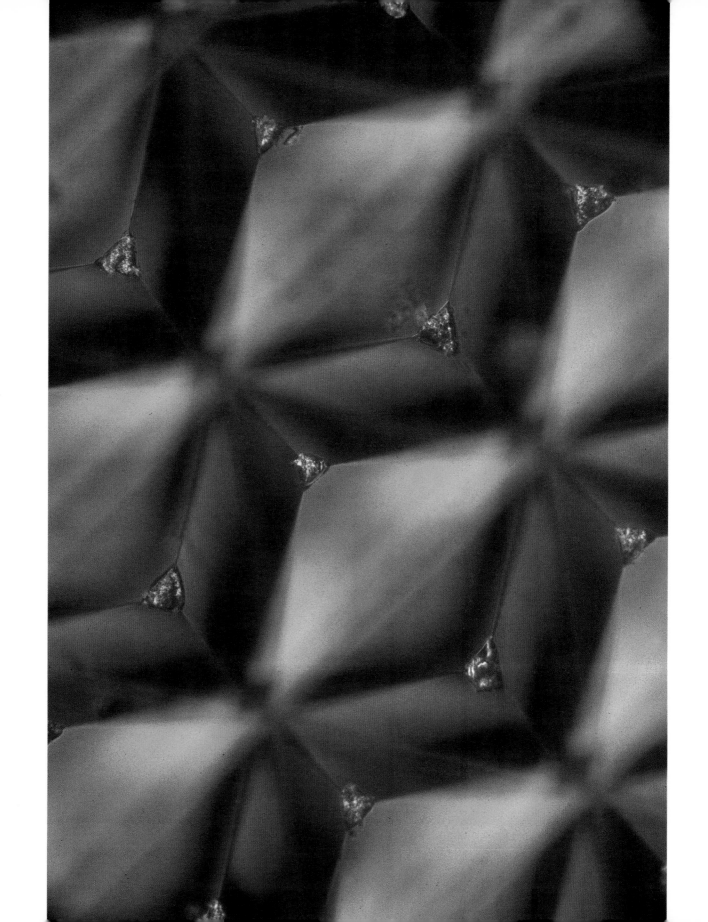

three perpendicular reflective walls. When light falls into one of these pyramidal structures from a broad range of angles, it reflects from the three faces and is thrown back to its origin.

So long as the orientation of the reflecting planes is precisely controlled, the accuracy of the direction of reflection is surprising. A corner cube on the moon can be observed by a radar transmitter on earth; the radar signal the cube reflects is directed toward the transmitter that sent it.

The same phenomenon, with different details, is involved in the reflection of headlights from the eyes of animals at night, and in those flash pictures of families in which all eyes seem to glow the pink of an internally illumi-nated albino rabbit. In these structures light enters the eye and is reflected from the inside surface. Retroreflective elements are used for street signs, reflec-tive pavement markers, and protective reflective strips for bicyclists: all reflect light back to the driver of the car whose headlight is the source of illumination. In the past, street signs and reflective markers have been made of reflective beads that operate as "cats' eyes." The more sophisticated corner cube structures are now taking their place, as micromanufacturing and optics are being applied to consumer products.

In the antique game of pinball, there were bumpers that reflected the ball back in the direction from which it came: two rubber bands at right angles, anchored at the ends. The ball entered, bounced off one band and then the other, and rebounded toward its origin in a retroreflective flight. When light encounters a simple mirror at an angle, it ricochets from the surface at that same angle but in the opposite direction—away from its source. Retroreflecting ("reflecting back") structures reflect light *back* toward its source.

These retroreflecting structures are an array of corner cubes molded into the back side of a sheet of plastic. A "corner cube" is half of a cube—

retroreflectors

sun, reflected

PREVIOUS PAGES

Looking at a reflection is looking into a world that does not exist. Most disorienting, if you think about it for even a moment. When the reflection is in a mirror, it is at least recognizable from its context—from the frame, from the discontinuities at the edge. In a window—from the surface of transparent glass—it is more disturbing. *There* is the reflection in the pane, and *there* is the reality behind the pane. Or is it the other way around? Which is which? How is it possible that what your eyes report so faithfully to you—the sun *there*, the trees *there*—can be a complete lie?

We understand how it happens. The surface of glass *is* reflective, although weakly. The light causes the electrons in the atoms at the surface of the window to oscillate; these oscillations re-emit light in all directions. In certain directions, the reflected light from the many atoms recombines and faithfully replicates the original, but with the apparent source appearing on the opposite side of the pane from the real source. All sensible and rational, but the eyes don't know physics, and the brain has a hard time figuring it out.

There are the tricks we have learned to tell which is which. Here, common sense and the shadow at your feet suggest that the sun is not really setting inside a stranger's living room. The ripples and lumps on the surface of the old window panes give clues. But this puzzle is one for the mind to work out. As far as the eyes are concerned, the sun behind the pane is as real as the one in front of it.

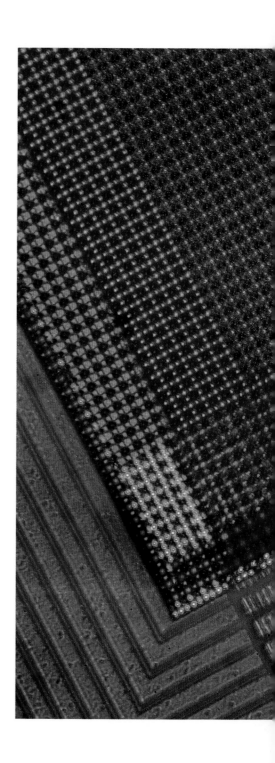

32

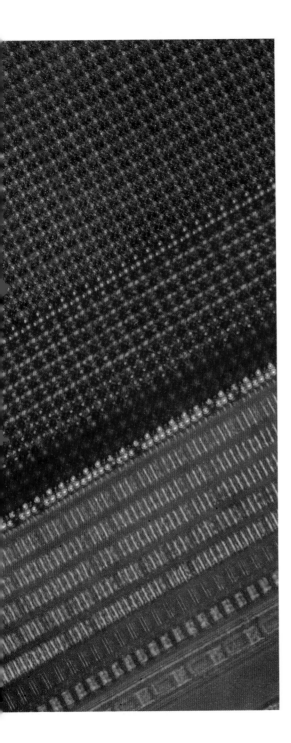

Tasting, smelling, feeling, hearing—all convert signals outside us into perceptions inside us in ways that are remarkably complicated. "Hearing" involves a collection of cells that function as tuning forks in an extraordinarily sophisticated structure—the cochlea—that separates sounds by frequency. "Smelling" and "tasting" are the recognition of particular molecules in the air or the saliva by complementary molecules in specialized cells in the nose and tongue.

"Seeing" requires a number of steps. In the first, photons pass through the lens of the eye and form an image on a gossamer sheet of cells—the retina—draped across its back. One kind of cell responds to the intensity of illumination with great sensitivity: these cells are color-blind, and see only black and white. They probably trace their way back in evolution to the earliest creatures to make the revolutionary discovery that a flicker in the level of light falling on them was sign of impending catastrophe. Those creatures with their primitive eyes sur-

vived because they twitched back into their crevice in the reef and abandoned their blind neighbors to be eaten. Other, more sophisticated cells—refinements in evolution—respond to colors. This collection of cells telegraphs the brain through the optic nerve. At the end, the brain performs its still-mysterious integration of position, intensity, color, and memory to tell us: "my child" or "fire" or "'Call me Ishmael.'"

What does a digital camera do? Something surprisingly similar in function to the first steps in vision. Light passes through the lens of the camera and forms an image on a collection of light-sensitive devices formed in a sheet of silicon. This collection—a "focal plane array"—is the camera's retina; the individual devices also record intensity, albeit by processes completely different from those used by the cells of the retina. These devices in the focal plane array see only black and white; color sensitivity comes from a color filter that allows only one color to illuminate each device.

33

focal plane
array

silicon, etched by light

We know light best in its diluted form: a gentle rain of photons falling from the sun that illuminates and warms. More concentrated, light is a furnace and a terror.

The surface here is silicon, the element that forms the basis for the micro-processors used in computers. It is very hard, very strong, and very brittle. Machining silicon—carving it into pieces—and seasoning the pieces by adding traces of boron, phosphorus, and arsenic to change its electronic properties is the art of microelectronics fabrication. Without this art, making transistors would be impossible.

Many tools are used to shape and transform silicon. A new one, made possible by the development of lasers, is light. When the focused spot of intense light from a laser traversed the surface of this flat piece of silicon, lanes of parallel, closely spaced lines—like those left by combines moving through wheat—were burned into the surface.

Silicon melts only at a very high temperature and boils at an even higher one. It is also unreactive. Here, the stream of photons was intense, and each photon that was absorbed deposited its energy as heat. The hot silicon combined with a reactive gas, chlorine, present in the atmosphere over its surface. The product of this reaction, silicon tetrachloride, boiled away, leaving the grooved lines in its place. The numbers are reference marks left in the untouched part of the surface.

Microelectronic devices built from silicon are everywhere and invisible: in computers, electronic ignitions of cars, television sets, telephone systems, and elevator controls.

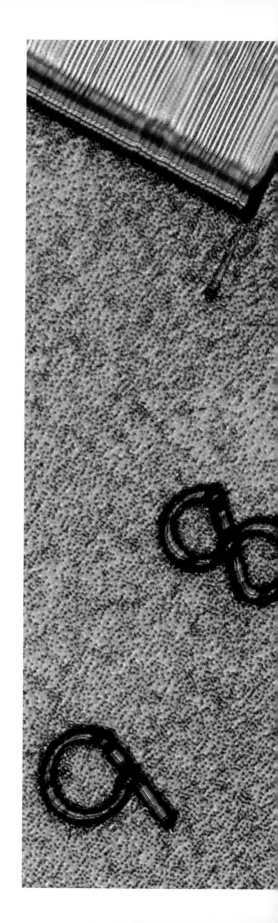

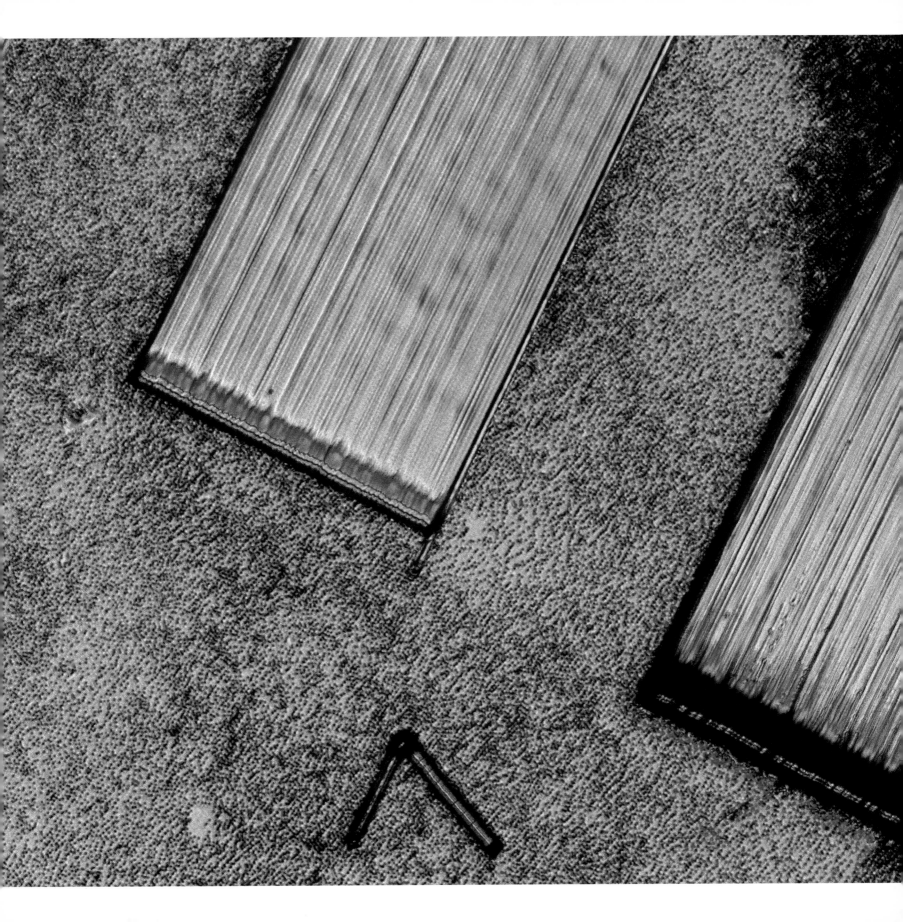

form

We are drawn to surfaces
for their reality. Surfaces define form;
form reflects utility
and history.

Pick up a baby, an apple, a knife; their surfaces tell you their shape, and their shape tells you what you have.

We are drawn to surfaces for their reality. Surfaces define form; form reflects utility and history. Surfaces are especially important for the forms of fluids and small things. Atoms exposed at the surface of any material are less stable — less comfortable — than those buried in the interior: liquids and solids try to reshape themselves to minimize their exposed surfaces. More of a small object than a large one is surface, and the desires of the surface atoms to escape exposure influence the form and performance of small things. It is easy for a liquid to contract its surface, and minimization of liquid surfaces shapes drops and films. Solids cannot change shape, and must use other strategies to cover their exposed surfaces.

A gear, with teeth transmitting force; a transistor, with hidden channels metering electrons; a knife, with its cutting edge — for each, function follows form. Snowflake and raindrop are water shaped by natural processes: their forms are their faces, and tell their stories. Form emerges from design, fabrication, and manufacture. Form also grows from natural forces, without our intervention.

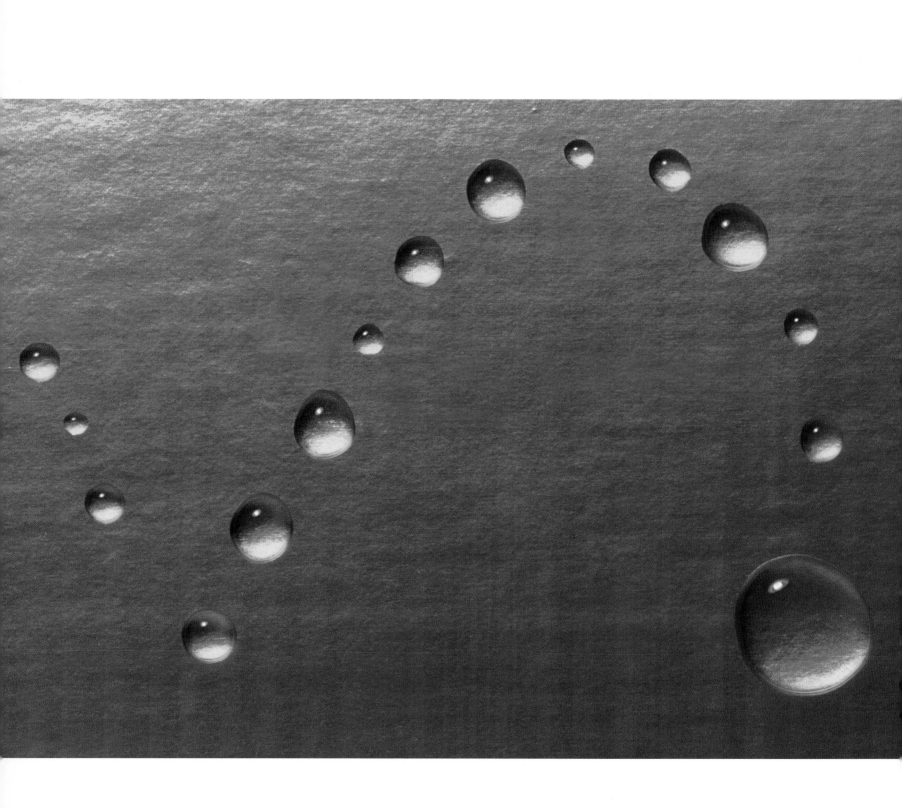

drops of water

Starlings scatter to feed, and clump
to sleep. They are attracted to the fields
or to one another.

Water draining from a freshly washed
dish sheets or beads: it forms either
a thin continuous film, or separated
drops. The surface of water exposed to
air tries to contract the drop; the
surface of the dish tries to expand it.
Competition between the single
layer of water molecules at the surface
of the drop and the single layer
of molecules at the surface of the dish
determines the shape the water takes.
Either the molecules of water are
attracted by the surface of the dish
more than to one another, or they are
attracted to one another more than
by the dish. The shape of a drop
is established by its outermost single
layer of molecules—it is visible
to the eye. Remarkable. The shape
of a drop is one of the few places in
nature where we can see the action
of single layers of molecules.

It is straightforward to explain beading
of drops: molecules in the surface
of the water attract one another more
than they are attracted to the surface of
the dish. It is more difficult to explain
the pattern of separated drops. As
the water slides from the dish, it initially forms a continuous sheet many
thousands of molecules thick. As this
sheet of water becomes thinner,
the tendency of its surface to contract
becomes greater. Small ripples on
the surface of the water, or small spots
of different wettability on the dish,
cause the sheet to break apart.

Whether a film of liquid remains continuous or breaks into drops determines
the efficiency of condensers in power
plants, the coverage of surfaces by
paint, the condensation of mist on ski
goggles, and the comfort of contact
lenses. The wetting and spreading of
liquids influence performance wherever
a liquid contacts a solid surface.

ice crystals

When something interesting happens on the street, a clump of people gathers to watch. The clump itself becomes an attraction, and passers-by attach themselves to its periphery. The crowd first grows irregularly, then it becomes continuous as more people push in. The formation of ice on the bathroom window in winter is similar to the growth of crowds.

When it is cold enough outside, the temperature of the inside surface of the window falls below freezing. Water from the air condenses, and then freezes. The first, tiny crystals of ice form with difficulty: their volume is small relative to their surface, and they are less stable than larger crystals. They require a starting point—a particle of dust or an especially cold spot—around which to organize themselves.

Once the initial crystals—the crystal nuclei—have formed, their growth occurs smoothly. Water molecules arrive from the air and attach to the outside of the expanding crystal. Because the surface of the glass is colder than the surface of the ice, crystal growth follows the surface of the glass. Here, the patterns of the ice, illuminated in the golden light of a setting sun, trace the patterns of growth of the crystals.

The two major stages in the formation of crystals—nucleation and growth—determine their sizes, shapes, and internal structures. These characteristics are widely important. The flow of powdered drugs into capsules, the dusting of sugar on cookies, the growth of concrete, the strength of steel, the performance of silicon-derived transistors—all are influenced by the properties of crystals.

40

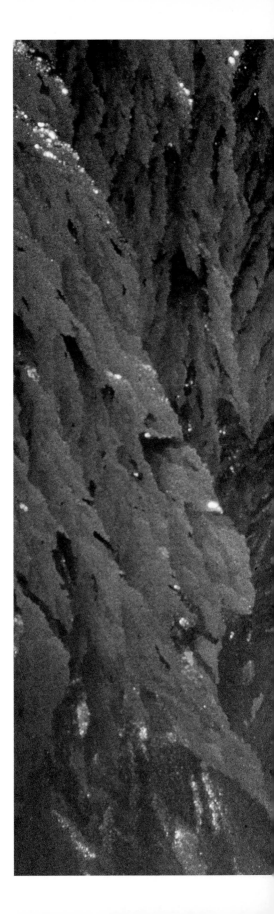

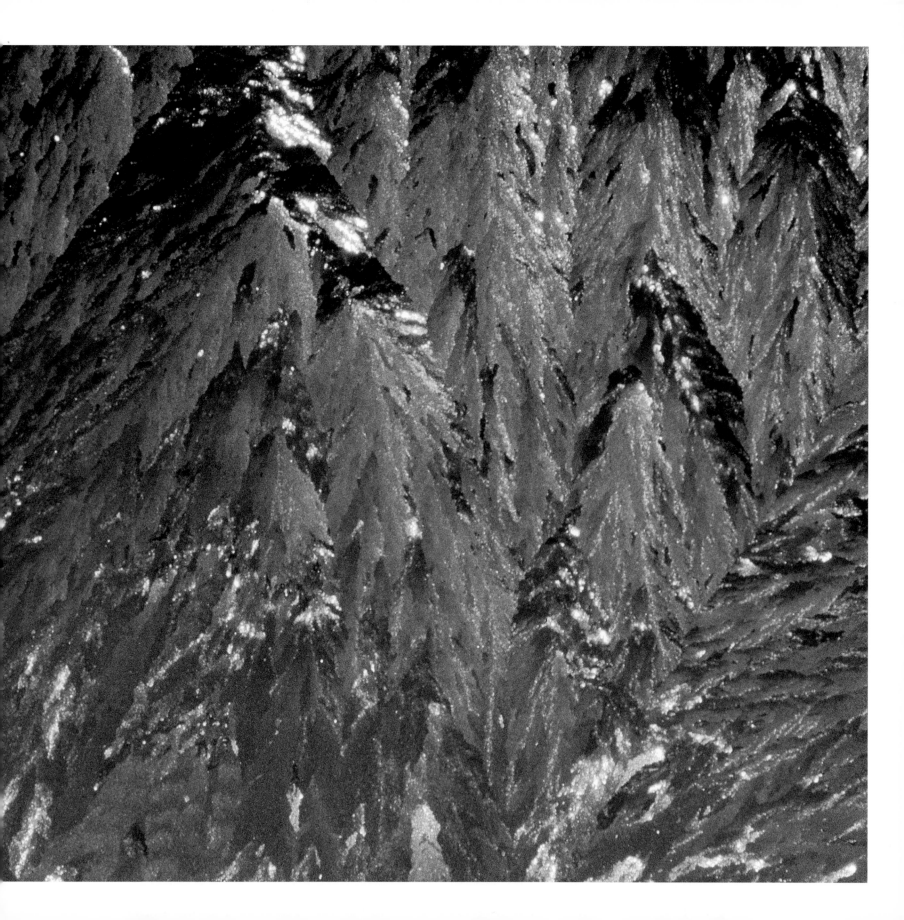

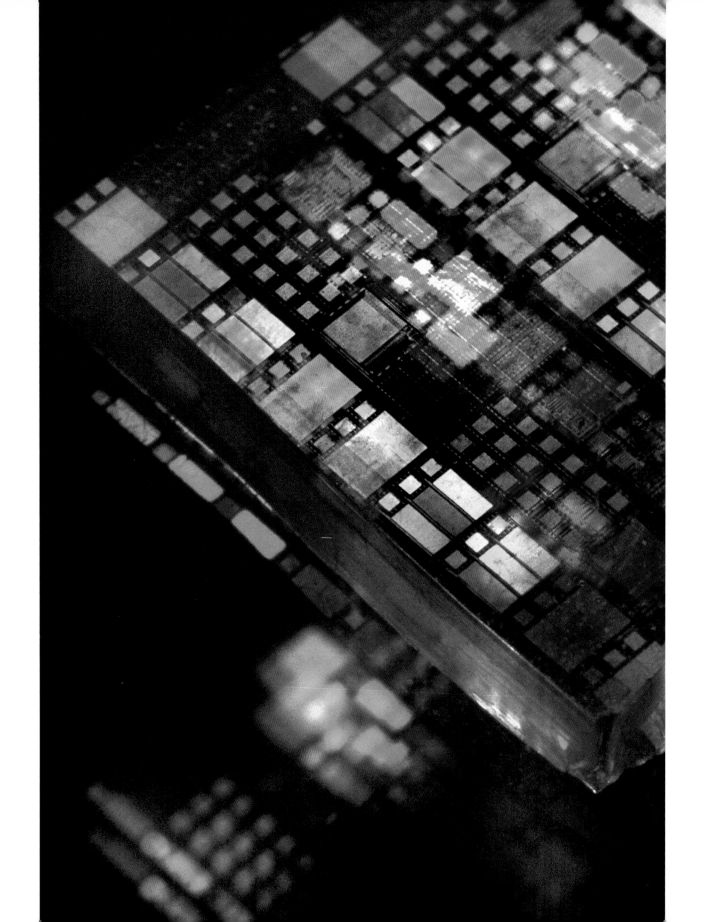

Making things small has become our passion. Medieval popes raised cathedrals toward Heaven that amazed the eye and lifted the soul. We build instead toward invisibility: we cannot even *see* our "cathedrals," our microprocessors and genetically engineered bacteria.

The most elegant of our miniature constructions span a few thousand atoms—a few hundred nanometers, a thousand times smaller than the period at the end of this sentence. They can be carved down from crystals or built up from atoms. Building up from atoms is still difficult; techniques for carving are more developed. Those who sculpt electronic complexities into crystalline silicon first photograph their designs on its surface, and then develop the images into structures the electrons can understand, using disciplined applications of etches, ions, and reactive gasses.

The structures made in the microworld of electronics have shrunk inexorably, and so has the size of the light used to image their designs. Photolithography, the photographic technique used, cannot be used to make much smaller images than it does now. When structures shrink to just half the current smallest sizes, the cameras will fail: no lenses are transparent to the required wavelengths of light—the colors between ultraviolet and X ray. New methods of building microscopic structures use electrons, ions, or X rays instead of light. A less conventional method is to use a rubber stamp to print patterns.

This stamp is approximately 2 cm across. The depth of the relief of its surface features is approximately 2 microns, and the narrowest of the lines it prints is less than one micron wide. The stamp is transparent to visible light: it looks like a piece of clear glass, although it is rubbery. The "paper" for this stamp is a gold mirror: a gold film 100 atoms thick, supported on flat silicon or glass. The ink is a colorless chemical that adheres strongly to the surface of the gold; the trace it leaves is exactly one molecule thick. The image shows the stamp and its reflection in a mirror. The colors diffracting from the surface of the stamp indicate that the patterns engraved into it are regular and have micron-scale features.

Microprinting is being developed as a tool in microfabrication. Although the patterns formed by these stamps are not visible to the eye, they can direct the attachment of mammalian cells to the surface or control the etching of a gold film. Stamping a film of gold with a pattern, and then etching it, generates small gold wires and mirrors for optical sensors and microanalytical equipment. The gold patterns can themselves be used to direct the etching of silicon.

stamp for microprinting

nonwetting
surfaces

How does a plant keep its leaves dry when it rains? Plants also need to keep dry; a coating of water hinders the flow of carbon dioxide into the plant and oxygen out of it, and slows photosynthesis.

We know how animals keep dry. Sheep—not otherwise the most clever of animals, but ones that live in cold, wet climates—have blazed this trail. Sheep cover themselves with thick coats of wool, and then coat the wool with grease. The greasy wool keeps them dry: on it, water beads and falls off. Of course wool keeps sheep warm, but a key part of being warm is being dry. Think of a sheep as a sweater that has been lightly spritzed in cooking fat. With legs.

But why does this strategy work, and what does it have to do with plants? Water has a high surface tension: the molecules on its surface really, really do not want to be there—they want to be inside the drop. To minimize its

surface, a drop of water normally adopts the shape—a sphere—that gives the smallest area of surface for a given volume. A spherical raindrop is an example. Water only spreads into a flattened shape when it is forced to do so by interactions with the surface on which it rests, or when it has been tricked (for example, by the addition of detergent).

So, a strategy for staying dry is to develop greasy wool or fur or hair (or something functionally like it); on such a surface, the water will bead and roll off, rather than soaking in. It works with sheep and other animals, and this leaf shows how it is done by plants. The leaf has a sparse growth of hairs on its surface (because warmth is less of an issue with a plant than a sheep, "sparse" is sufficient); these hairs are covered with a waxy film. A drop of water resting on this surface finds nothing to like. It rests obliviously on the tips of a few oily spines, and is mostly surrounded

by air. It responds by rolling up into a sphere, which, like any ball, rolls off the surface of the leaf when it moves in the wind.

One master of this strategy is the lotus plant, whose combination of rough, textured surface (more warts than hairs, but they serve the same function) coated with wax makes it most inhospitable toward drops of water; so much so that the combination of roughness and hydrophobicity is now called the "lotus effect," and is being developed to control adhesion of dirt to house paint and rain to windshields.

This leaf rests on a grid that illustrates the principle, although on a scale much larger than the hairs on the leaf. The surface of the grid is approximately half air. This grid, if coated with Teflon or candle wax, would also repel water, but air is even waxier than wax in its indifference to water.

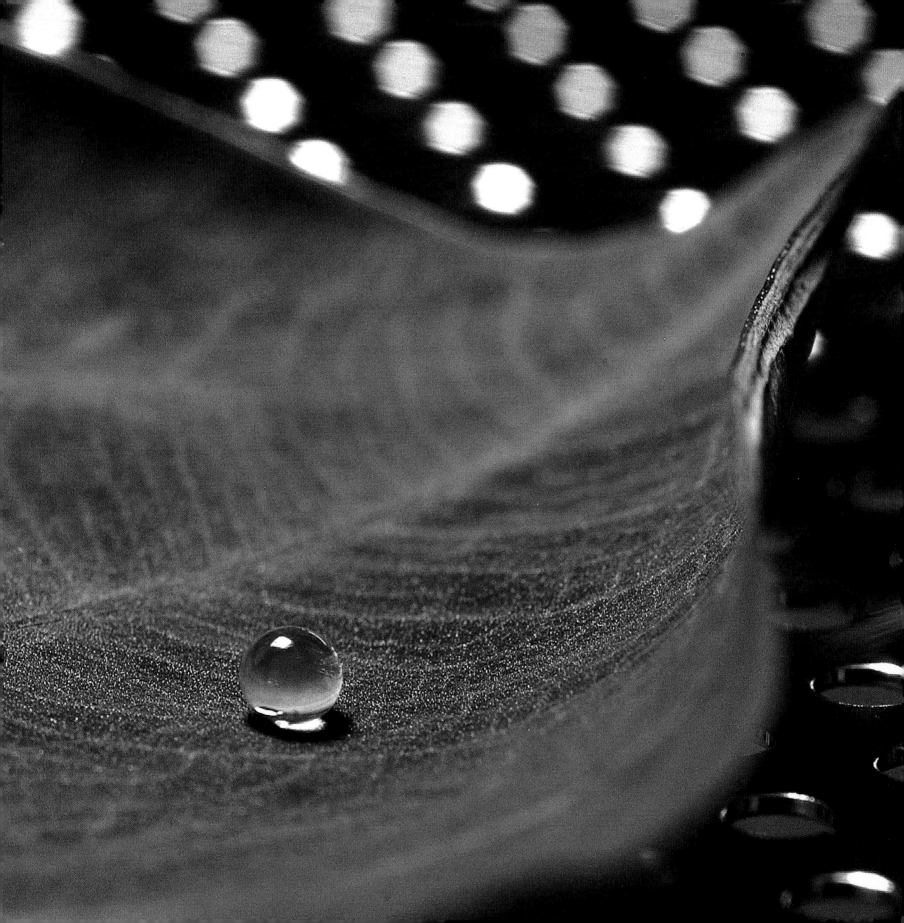

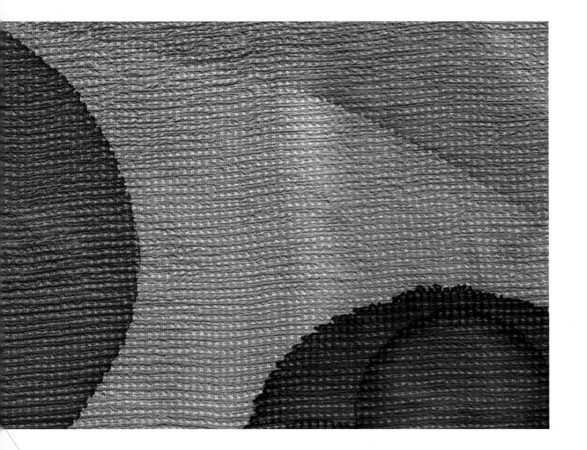

Molecules—like ants, lemmings, herring, people—are happiest when surrounded by their own kind. Molecules exposed at surfaces are always anxious: on one side they are comfortably surrounded by kith and kin; on the other is empty space. Molecules in liquids deal with their agoraphobia easily: they cluster together and form compact drops. Molecules in solids cannot move and cannot cluster. When a solid finds a liquid, a tug of war follows: the solid tries to hide its nakedness by pulling the liquid over it; the liquid resists, and does its best to remain a modest, compact drop.

Wine spreading on a linen napkin, blood staining a cotton bandage, ink blotting paper—all are contests the liquid is losing. The same conflict, played under more complex rules, describes oil sticking to rock in underground reservoirs, tears wetting contact lenses, glue sticking to the flap of an envelope. Matter does not like to exist bare at surfaces: it will try to cover itself with anything within reach.

inks bleeding on fabric

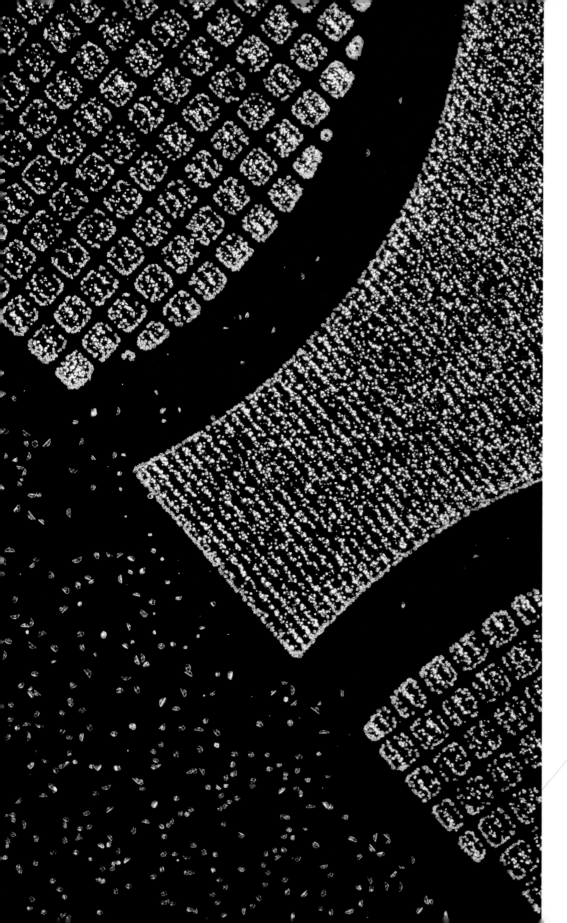

patterned crystals

If you are a mollusk (as many of our distant fore-bears were), you have a central problem. You are soft and tasty, and not good at sprinting away from something hungry (as almost everything is). How do you survive long enough to have offspring? The best answer is to build an armored structure—a shell—around yourself. Engineering is the final answer.

Fine. But a shell of what? The ocean in which you live is water—not great for building armor. But dissolved in the water are many ions. Although, as a mollusk, the periodic table was probably never your thing, evolutionary chemistry—trying every possibility randomly—has taught you that concentrating certain of these ions from the sea causes them to form crystals; you can use these crystals for armor! The best compromise of availability and strength seems to be calcium and carbonate ions. We recognize this combination—calcium carbonate—as limestone, chalk, or marble, and also name it "calcite." Crystals of calcite are very brittle. Oysters learned how to form it in thin sheets, with even thinner layers of organic glue between them. This combination is astonishingly tough and durable.

These three patterned arrays comprise small crystals of calcite grown in a laboratory. The surface of the lines and squares promotes nucleation; the spaces between them suppress it. Examining the crystals suggests how the oyster might have started to grow its house. How it finishes is a secret it still has not yet fully shared with us. Probably wisely. We also eat oysters.

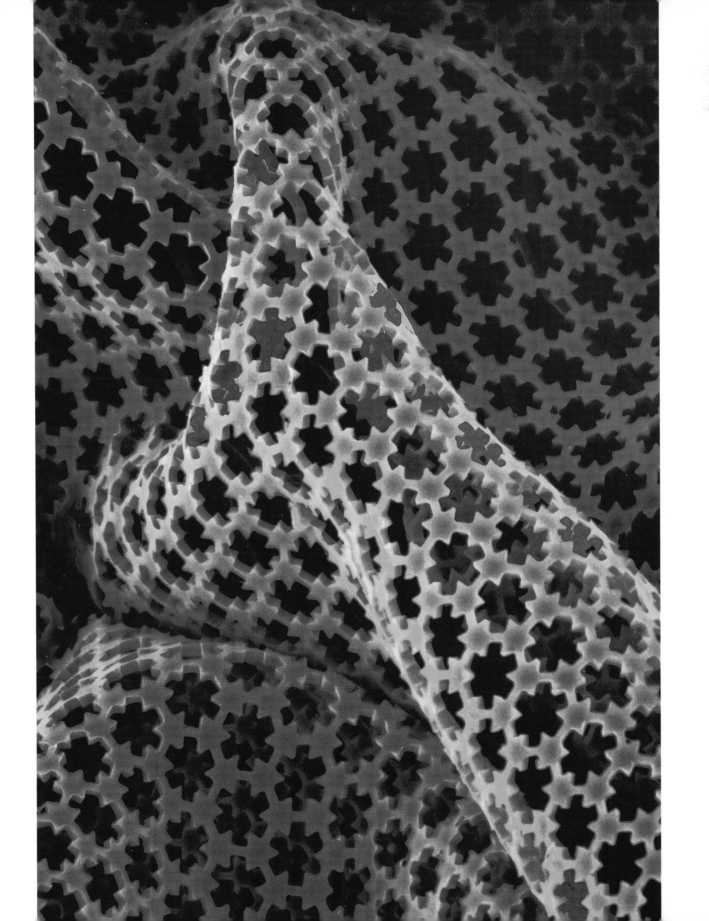

48

molded plastic microfabric

The spider and the Belgian lacemaker would be envious. Each makes a fabric of admirable regularity and delicacy: the one spins to catch the fly; the other knots to catch the fancy. This lattice-work fabric is more delicate than either; its use may be to catch light.

Its narrowest links are a micron across—one twenty-thousandth of an inch. The finest threads of spider silk are 10 times thicker; the finest lacemaker's thread a ship's hawser by comparison: 100 times larger. To make such delicacy requires a new craft; the older methods do not work. The spider's spinning apparatus is too sophisticated for us to mimic. The spider also works one thread at a time: she cannot spin simultaneously at many points, and her life would be too short to make the enormous number of connections (3 million per square centimeter) in this fabric. As for lace: there are no lacemaker's hooks and needles at this scale, and no human hand will ever be steady enough to do the work.

This fabric is made of molded plastic, not spun or woven: it is like a perforated rubber doormat, reduced in size 10,000 times. The mold was a network of capillaries formed by placing a block of rubber, with a surface textured in the pattern of the fabric, in contact with glass. A liquid precursor of the plastic to be molded, placed at one end of the network, filled it by capillary wetting. The liquid was transformed to solid by shining light on it, and the mold was peeled from the support.

The dimensions of this structure are those of crude microelectronic devices; the finest wires used in high-performance devices are ten times smaller still. One possible use for this type of fabric is as a pattern for photolithography, where it would have the advantage that it could be used once and then thrown away.

49

self-assembled structure

A biological cell is its own instruction book: it assembles itself. Its complexity comes from the organization of the molecules that make it up. These molecules organize themselves; no force external to the cell plays a role.

The self-assembly of a cell is one of the most marvelous feats in nature. Its molecules sense one another by feel: complementary shapes nuzzle and slide together into pairs, groups, and multitudes. It is a society of molecules—a city-state. It feeds, defends, and repairs itself, and it builds the components that allow it to replicate and form colonies.

By comparison, these cross-shaped pieces are performing a stupid pet trick, like turtles that have been taught to form a figure. They are floating at the surface of a liquid. By design, the liquid wets their bottoms and flat sides. As with soap films and drops of liquid, the surface of the liquid tries to shrink; as it does, the blocks slide together into a regular array. Minimization of the surface area of the liquid pulls the assembly together.

Although these pieces are large enough to be seen unaided, the strategy that caused them to assemble into a regular structure should also work for much smaller units. One future approach to the fabrication of complex structures for information technology—computer memories and microprocessors—might involve making simple components and allowing them to self-assemble into complex constructions. These structures are now manufactured in layers— millions of transistors at a time. The processes used are becoming intractably difficult and expensive, as the computers that use them become unimaginably complex. Any strategy that might provide a simpler alternative is interesting for future computer design. Ideas derived from biology were once believed too unfamiliar and too risky to be considered. Now, there is no choice but to consider the most imaginative strategies.

50

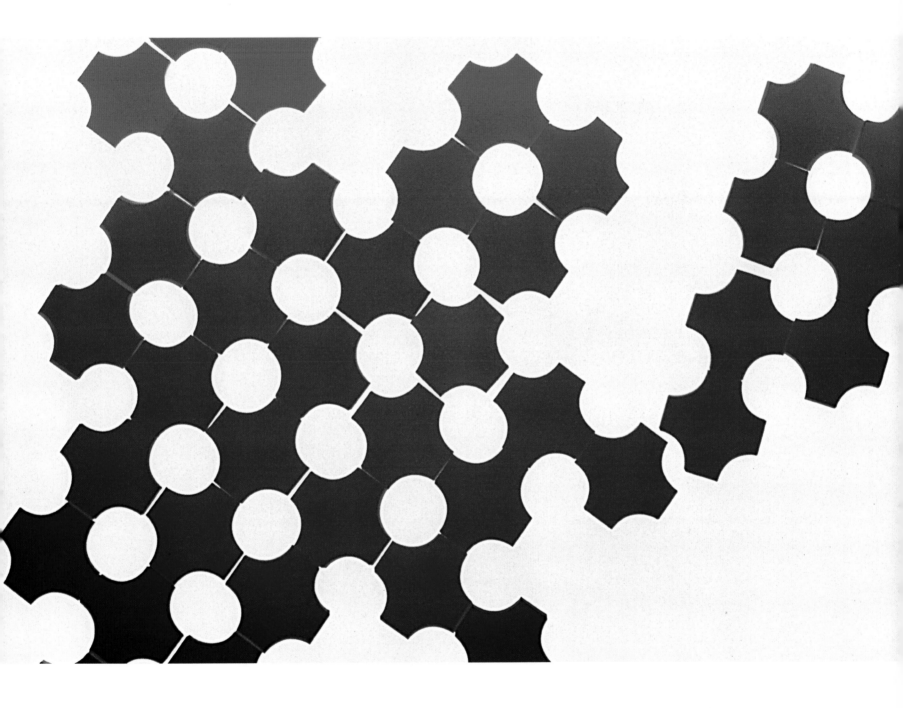

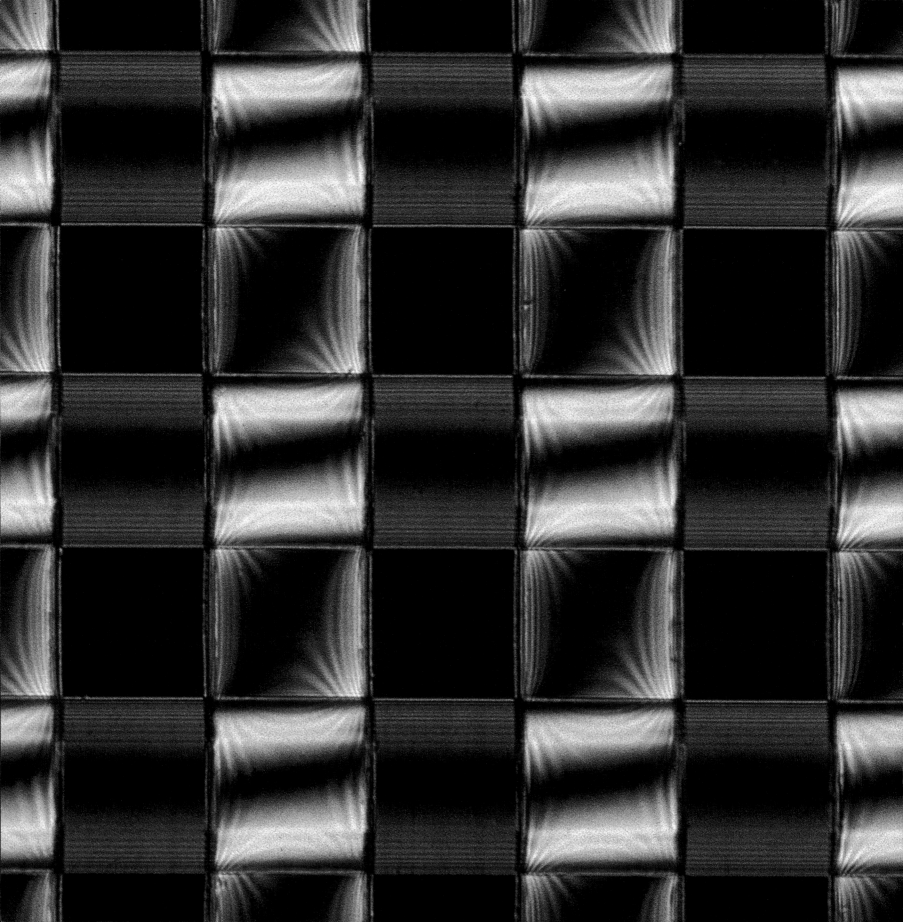

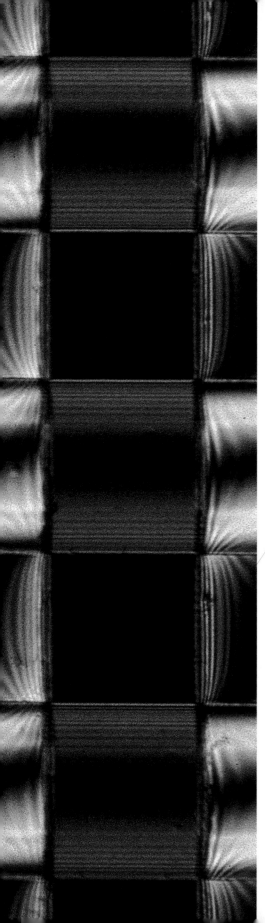

Painting the outside of a house—
white on the clapboard walls, blue on
the trim—is detail work: slow and
expensive. Getting the borders on the
trim sharp, and not smearing the
new white with blue while edging, is
a maddening job. It would be simpler
if the paints knew their places:
blue *only* on the trim, and white *only*
on the clapboard.

This grid is made of silver wires, each
covered with a thin transparent
film of electrical insulation. The insulation was painted on the wires by a
self-assembling process: the paint knew
where to go.

Making this structure started with
printing a pattern of two different
kinds of molecules on the surface of
a continuous silver film. The entire

coated silver wires

patterned structure was dipped into the
transparent insulating paint, a liquid
plastic. The plastic wetted (and
stuck to) only the parts of the surface
covered by one of the printed patterns;
this pattern designated the areas
intended to become wires. After the
plastic was cured to a solid, the silver
surfaces not covered by plastic were
removed by etching. Voilà! A set
of parallel wires covered with insulation. Repetition of the process with the
second pattern perpendicular to
the first made the second set of wires.

Self-assembly has the great charm that
it works with very small structures.
Finding a paintbrush to spread liquid
plastic on wires 50 microns wide
would not be simple: it is much easier
to let the surface and the liquid work
it out between themselves.

53

There is no central planning bureau that oversees evolution. The enormous complexity of even the simplest of living organisms was developed by throwing dice. Biology works by random trials.

It is disconcerting how much more sophisticated the products are from this mindless system than from the research programs that science so thoughtfully plans; it is also remarkable how slowly biology walks the meandering pathway of evolution. We cannot begin to equal the subtlety of a bacterium in our designs; we also cannot try random experiments for a few hundred million years to see if something good comes of them.

multiple
experiments

These spots represent a primitive effort to join random and planned experiments. The hope is for a fortunate combination of unanticipated successes from random experimentation, guided by some planning.

A cook tries recipes for cookies a batch at a time, one recipe differing from another by a pinch of this or that, each taking time to mix and bake. Here the "cookies" are very small, and each is made from a separate mixture of ingredients, so the cook — the scientist — can try many combinations of ingredients simultaneously. The ordered array helps keep the recipes separate.

Each small rectangle on this laboratory cookie sheet is a mixture of chemical elements similar to those present in so-called high-temperature superconductors. These remarkable materials conduct electricity without losses due to electrical resistance, and they have caused us to rethink what might be possible for the efficient transmission of energy. The high-temperature superconductors function at tempera-

tures much above those required for conventional superconductors, but they still operate at temperatures below that of liquid nitrogen — much too far below room temperature to be broadly useful. New superconductors, perhaps operating closer to room temperature, might be formed from an enormous range of different elements: there is no theory yet to guide selection and composition. Since we do not know why one combination works and another does not, we try random mixtures. One key to success is to test large numbers of combinations efficiently.

The strategy of trying multitudes of experiments in miniature is new. It is being widely practiced in the development of new drugs. Analogs of this strategy, which is often called combinatorial synthesis, are widely used to select single microorganisms from large numbers that have been forced to mutate in the hope that they will develop useful properties. Randomized, parallel experiments may increase the efficiency of research by allowing one researcher to conduct multitudes of experiments at the same time, and by decreasing the expense of disposing of completed experiments.

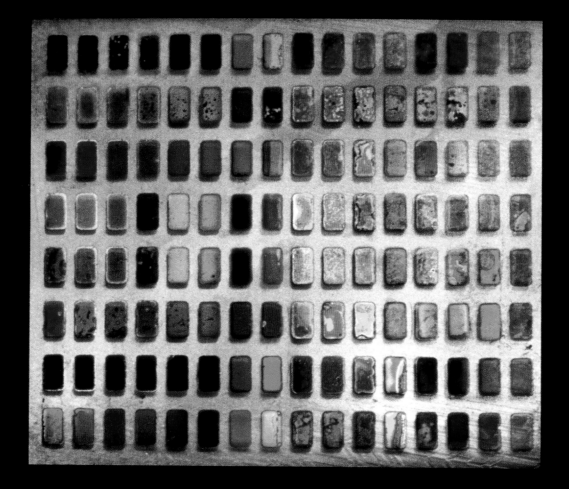

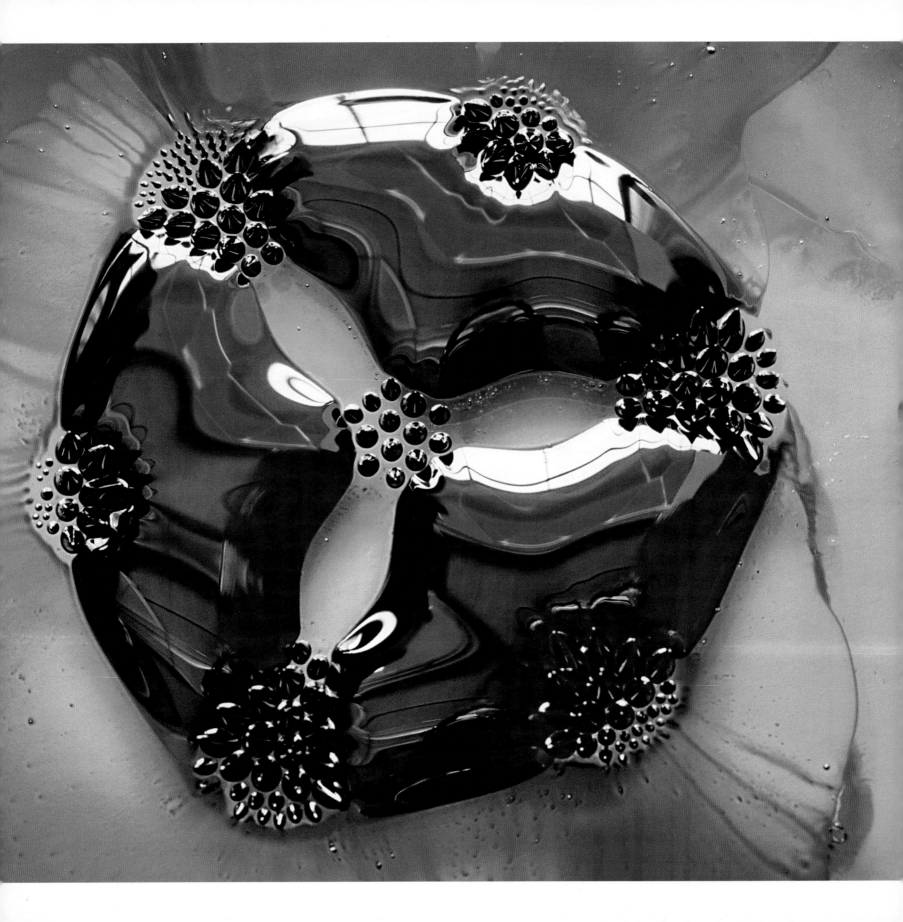

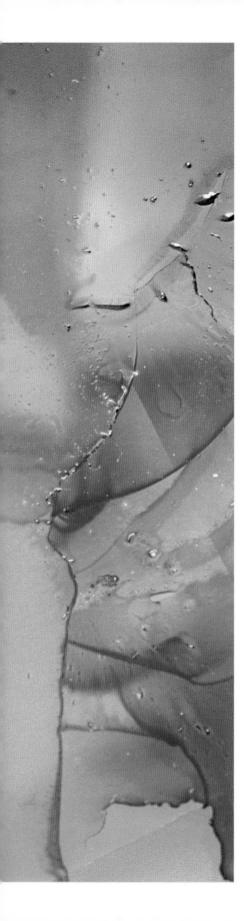

Pity the gryphon, the mermaid, the silkie, the chimera: creatures assembled of incompatible parts, with uncertain allegiances and troubled identities. When nature calls, which nature is it? When instinct beckons, approach or flee?

A ferrofluid is a gryphon in the world of materials: part liquid, part magnet. It is prepared by grinding magnetite—the magnetic lodestone—in an oil. The grinding must be "just enough." If the particles of magnetite are too large, they remember who and what they were and behave like a fine magnetic powder, clumping and settling rapidly from the oil. If they are too small, they no longer show any of the wonderful cooperation between groups of atoms that is required for magnetism. If they are just the right size—if they are small enough that they are not so different in size and character from molecules of liquid, small enough that they have begun to lose their magnetic heritage, but still large enough that they again become fully magnetic when placed in a magnetic field—they develop a useful schizophrenia. Outside a magnetic field, they are non-

f e r r o f l u i d

magnetic liquids; in a magnetic field, they become magnetic. Grinding is carried out in the presence of soap, which coats the small particles with an oil-like surface film and makes it even more difficult to distinguish them from the oil. Properly reduced in size, and correctly coated, the particles remain dispersed and do not settle.

When placed in a magnetic field, the conflicting attractions of gravity, magnetism, and surface tension shape the ferrofluid. This drop of ferrofluid was placed on a glass sheet with yellow paper underneath for photographic contrast. Seven small magnets were placed below the paper. In regions of high magnetic field, the fluid broke into spikes, trying to imitate the way iron filings line up in columns in a magnetic field. In regions of lower magnetic field it remained a liquid, forming flat drops in a compromise between the siren call of gravity and its own cautious cohesion. The result is shapes seen nowhere else in nature.

The unique properties of a ferrofluid—a stable liquid that responds to magnetic attraction—make it useful in devices where fluid properties and resistance to gravity are needed, such as rotary seals in disk drives for computers, and dampers for high-performance audio speakers.

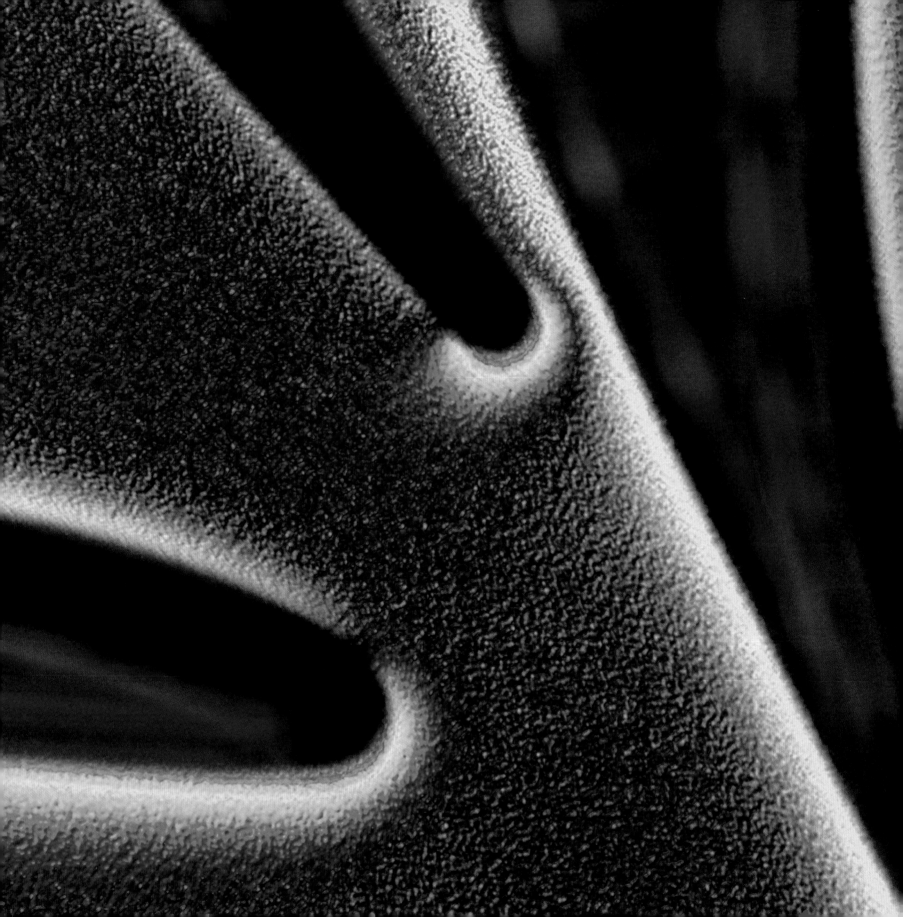

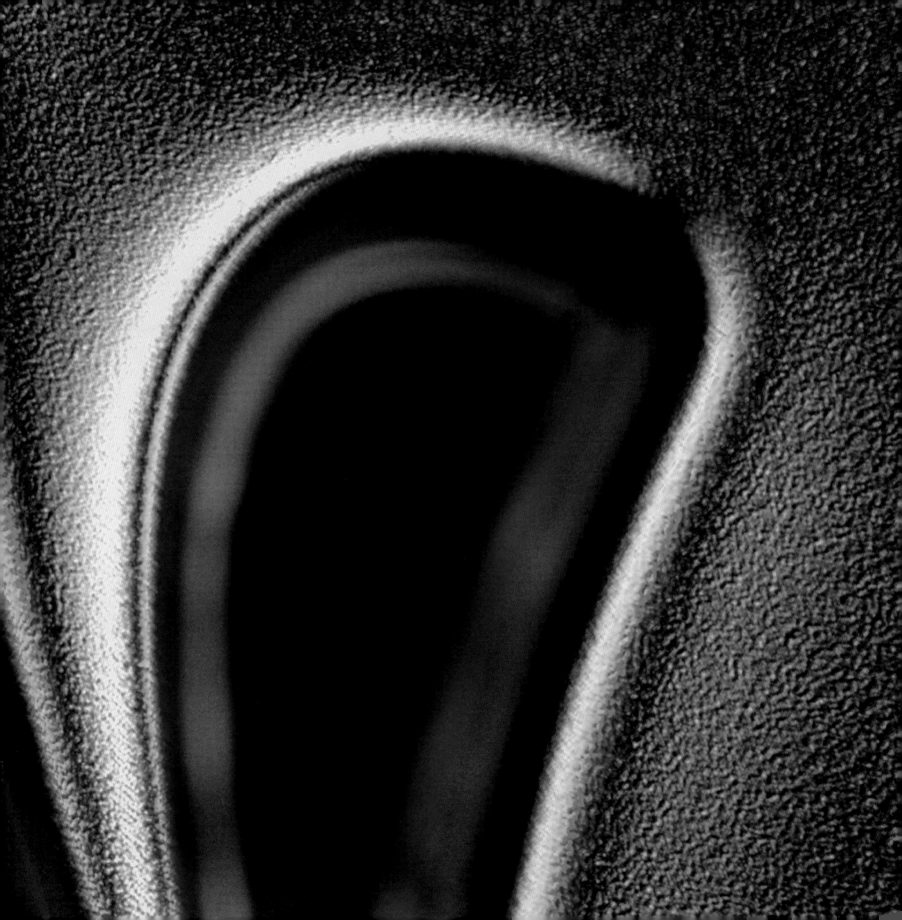

small machines

PREVIOUS PAGES

Mechanical things have a comforting size about them: automobiles, vacuum cleaners, and toaster ovens are the domestic animals of a world drifting out of contact with living things. They thump and chirp and bang in familiar ways; they pull and push and pump and carry. They are the sounds and activities of our barnyards.

Electronic things are disconcerting: small, and more arthropod than mammal. Their buzz is too shrill to hear; their tasks we might wish not to imagine; their forms are not instinctive. If they think, they think thoughts that are not ours. At least they do not move: we have them pinned in place.

This structure is a microelectromechanical system, a MEMS. It is a hybrid between a machine and a microelectronic device—a machine the size of a few transistors. MEMS are one of the next steps in the evolution toward the microscopic that is causing many of the tools of the world to shrink and disappear before our eyes. MEMS are new companions for us. Individually

they are too small to see. They are mouths chittering sounds we will never hear; eyelids blinking to observe worlds that would be dwarfed by our retina; scribes writing script that only electrons read.

The evolution of these very small machines has been remarkably rapid. This structure is carved from silicon using techniques developed to make microelectronic devices. It is a part of a small gas turbine. Although this device will only rotate when completed, and its cousins can bend and flex, all are still tied to the silicon chip from which we carved them. Future generations will move independently.

MEMS are now used to detect sudden decelerations in automobiles—as in a collision—and to set off airbags. They are parts of analytical systems that have the spatial resolution to locate single atoms. They are being developed for many new applications: arrays of micromirrors for large-panel display systems; tools for fabrication of microelectronic devices; even miniature inertial guidance systems.

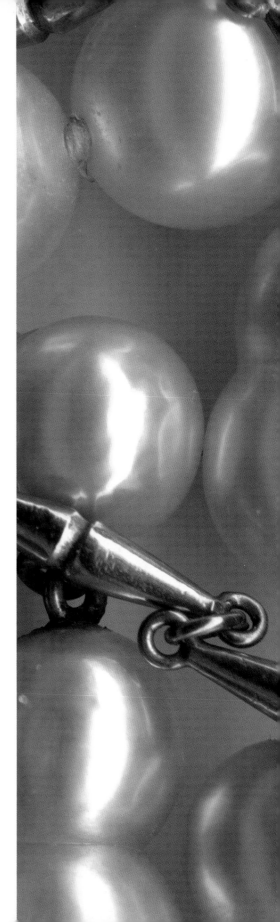

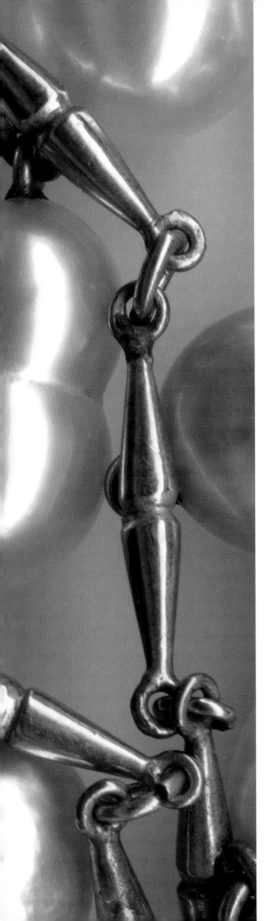

The lacquered surface of a Chinese tea chest was much prized by our grandmothers. Many coats of lacquer, patiently polished between coats, produced a hard, lustrous finish with no surface roughness. It required patience, effort, and time.

So it is with the oyster and the pearl. To the oyster, the seed for the pearl is a serious matter: gravel in the eye, a knife in the gut. The oyster has no fingers to pull the irritant out. Instead it carefully covers it in layers of glistening nacre. No Chinese chest was ever lacquered with greater care.

And why is it a shimmering, translucent white? It is the easiest thing for the oyster to do. A pearl in our hand is white for the same reason as snow in sunlight, or the tops of clouds, or frosted glass. It is made of many small transparent particles; those in pearl are held in place with transparent "glue." Each surface in this pearl—like each facet on a snowflake, or drop of mist in a cloud—reflects light. The light enters this maze of mirrors and bounces randomly from facet to facet: much of it ultimately bounces out again. The light at the center of a pearl must be as dim as that at the bottom of a heavy cloud.

The surface of the pearl does not irritate the tender membranes of the oyster. We have similar needs to make foreign objects compatible with our bodies: for implanted heart valves, blood vessels, and lenses, and for replacement veins and bones and teeth. Pearl and other solid structures that are innocuous to living tissue are models for the structures that will prop us up as we decrepitate.

61

pearls

order

order

forms the spine of our
efforts to measure, control,
and understand.

Order is repetition, regularity, symmetry, simplicity. It
forms the spine of our efforts to measure, control, and understand. We recognize and admire the simplest kinds of
order in nature — the intersecting planes of crystals and the
smooth curves of liquid drops. The order of more complex
structures — sea shells, zebras' stripes, opal — fixes our
attention. When there are no simple symmetries, order is
more difficult to perceive. The living cell is the grand example.

We pay in effort and energy to impose our order on the
natural tendency to disorder.

To keep sheep, you need fences; unfenced sheep spread, wandering after grass.

These colored squares are drops of water. They rest on a surface across which they would normally form circular drops and spread until they coalesced. Instead, they are performing unnatural acts: forming squares, staying in place, not mixing. What fences pen them? What grass attracts them?

The surface beneath the drops has been coated with a single layer of molecules of a type much loved by water (a hydrophilic surface). The water molecules cover it as completely as they can. But the surface has also been broken into square fields by painted stripes a single molecule thick of a different type of molecule, one the water avoids (a hydrophobic surface). The molecules of water crowd to the edges of these stripes, but they do not jump them.

The backs of sheep are often spray-painted to identify them; the colors show when part of a flock has found a gap in a fence and drifted into another field. Dyes added to these drops do the same for water.

The square drops are approximately 0.4 centimeters across; the stripes dividing them are 1 micron across and one-thousandth of a micron high. If the water molecules were sheep, the fence would be 3 sheep tall and 1,500 sheep across, and the sheep would be piled 1 million deep in the center of the field. It would be a peculiar form of agriculture.

Controlling the spreading of liquids on surfaces—the wetting of the surface by the liquid—is surprisingly important, not only in painting, printing, and gluing, but also in growing mammalian cells; in fabricating the myriad microelectronic devices that swarm around us in computers, cars, airplanes, and air conditioners; and in condensing steam in boilers.

64

square drops of water

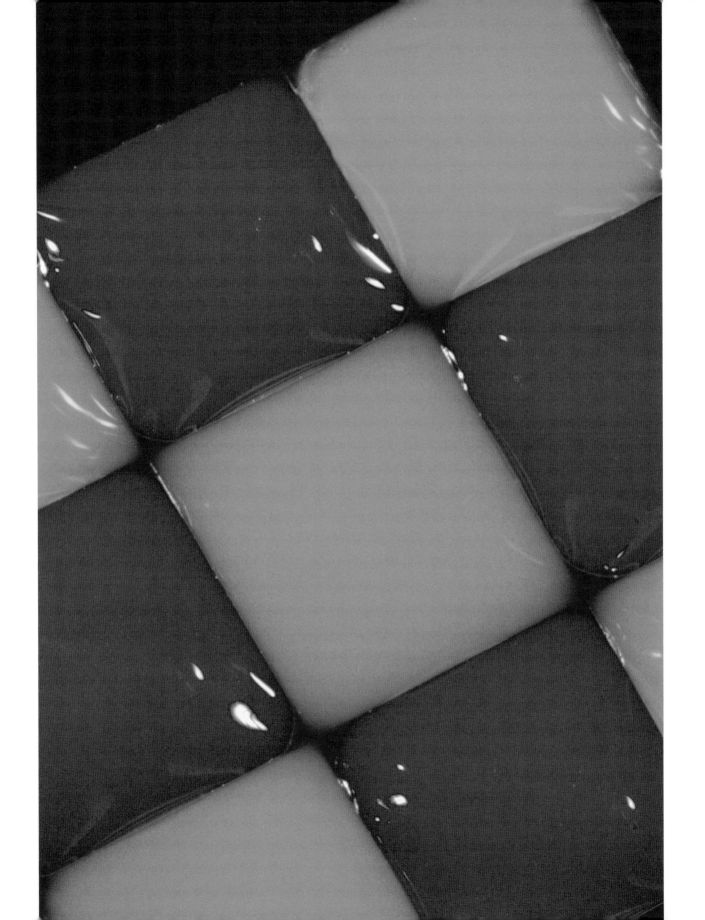

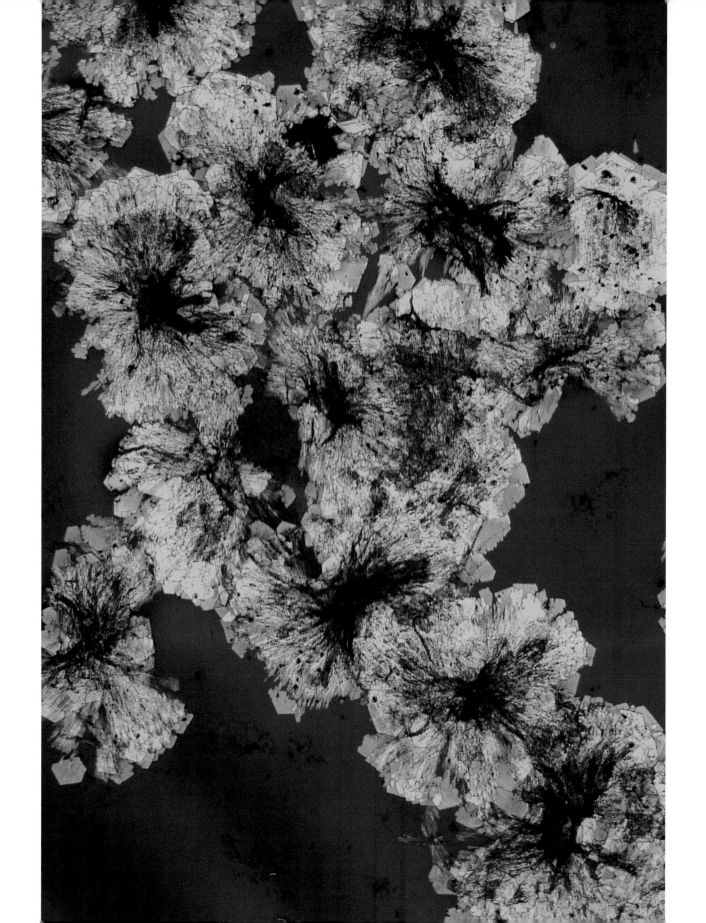

"Crystallization" is identical things arranging themselves in indistinguishable positions, and doing it by themselves. Molecules crystallize; TV sets and dollar bills do not. Being a multitude of identical things is not enough; the members of the throng must also attract each other, jostle into place, and settle into order.

Forming a crystal is delicate: the particles must all have the same shape; they must attract one another but maintain the proper arms-length distance; they must have enough freedom and enough motion to find their most comfortable position. For molecules, this balance of circumstances is common; for larger objects, it is not.

These arrays are crystals of small particles (approximately 50 nanometers in diameter) of cadmium selenide. Crystallization of these spherical particles occurred in sheets; in many areas, the sheets stacked. The crystals of spheres are like opals, except that the spheres are a tenth the size of those in opal. They are a step toward a new type of matter: crystalline arrays of small spheres of semiconductors— materials in which the movement of electrons is more fluid than in electrical insulators, but more hindered than in conductors.

Microelectronics is based on the unusual movement of electrons in semiconducting forms of silicon; in these arrays light, not bits, is what attracts us. The colors result from several phenomena: the yellow from the emission of light by the particles on illumination with ultraviolet light, the blue from the background.

Each spherical particle in a crystal sheet is a "quantum dot": a box in which an electron can move almost from wall to wall without striking an atom. Each "dot" is a drum whose electrons give off visible light when struck with the proper drumstick, ultraviolet light. The timbre and pitch of the drum can be tuned by weak connections to neighboring drums. Quantum objects—dots, wires, boxes, and other structures in which electrons, photons, and atoms are so constrained that they show new types of behavior—are an exotic new continent for physicists to explore. New rules for the behavior of matter will mean new technologies: perhaps more efficient memories and faster microprocessors for computers.

67

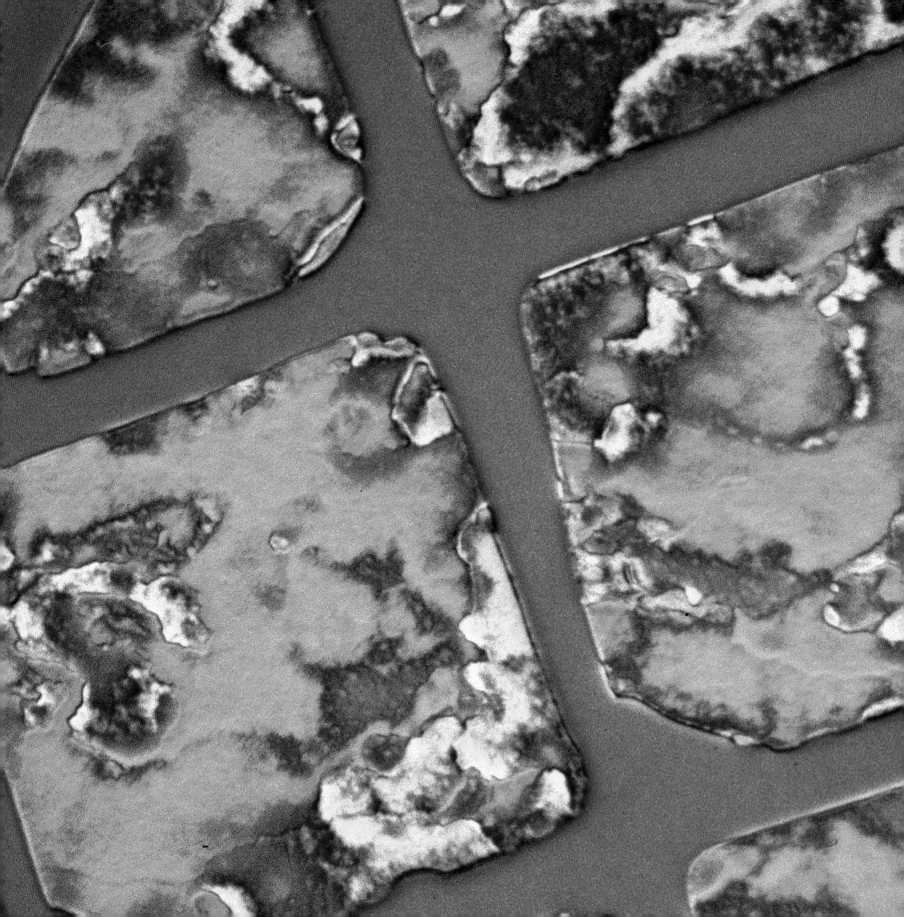

liquid crystal film

PREVIOUS PAGES

We know about gases, about liquids, about crystalline solids. They make up the adults of the society of molecules. The molecules of gases are loners, out of contact except for brief encounters. The molecules of liquids are condensed but disordered: a crowd, tightly packed, jostling and pushing. The molecules of crystals have the order of the military parade ground: identical elements in ordered, indistinguishable positions.

Then there are the adolescents—liquid crystals. Left alone, they swirl in apparent disorder, but they can obey an order as complicated as any followed by their elders. When commanded, they reluctantly line up. When left without external authority, they swirl again according to their own mysterious rules.

And what can they do? At best, their organization is loose; they are not suited for heavy-duty work. In the liquid crystal display of your watch or computer screen, they jostle light. In these devices, the liquid crystalline film is sandwiched between two electrical conductors: a metallic film and a transparent film of an electrically conducting metal oxide. When an electrical potential is applied between these electrodes the molecules line up, and light easily undulates through their loosely ordered ranks; these ordered regions are transparent and appear as black because a black backing shows through. Where there is no electrical potential—where the molecules are left without instruction—they collect in groups, and light passing through them bumps, stalls, changes course; the result is white, as the liquid crystalline film becomes "milky" and light scatters.

Here, a liquid crystal film was sandwiched between two surfaces. One instructed the molecules of liquid crystal to line up perpendicular to it. The second was patterned into a grid, with different regions sending opposing instructions: one region told the molecules to stand vertical, the second instructed them to lie flat. When the orders from both surfaces were the same, the molecules of liquid crystal lined up in an orderly way: those regions are brown. When the surfaces gave different instructions, the result was disorder: these regions are mottled. As always, with adolescents, conflicting instructions led to chaos.

Imagine a world in which there is only order: everything identical—in composition, form, orientation, and motion, as far as the eye can see. Move any distance in any direction, and you cannot tell that you have moved. Such order is deeply disorienting, if change is your guide, but deeply soothing if surprises distract.

With atomic order comes atomic serenity. In crystals, waves of light ripple over mirror-surfaced pools; electrons swim in ordered schools in oceans without tides. Crystals shelter phenomena—superconductivity, lasing—that require calm, coherence, order.

Crystals are essential for many areas of technology: silicon and gallium arsenide for fabrication of microelectronic devices; diamond for cutting tools; sapphire for lasers; quartz for timers for clocks.

70

man-made crystal

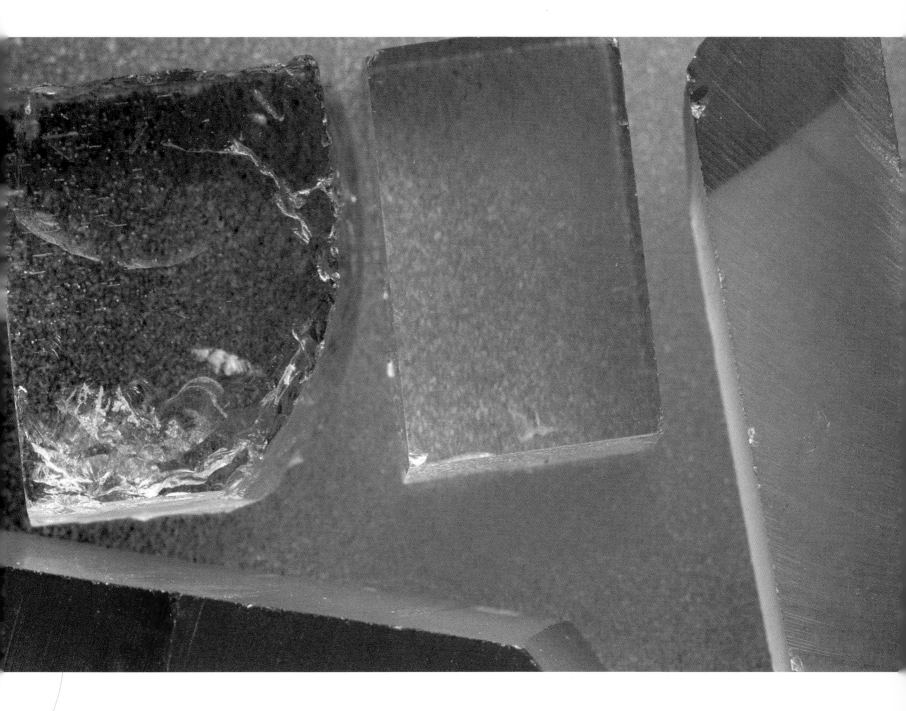

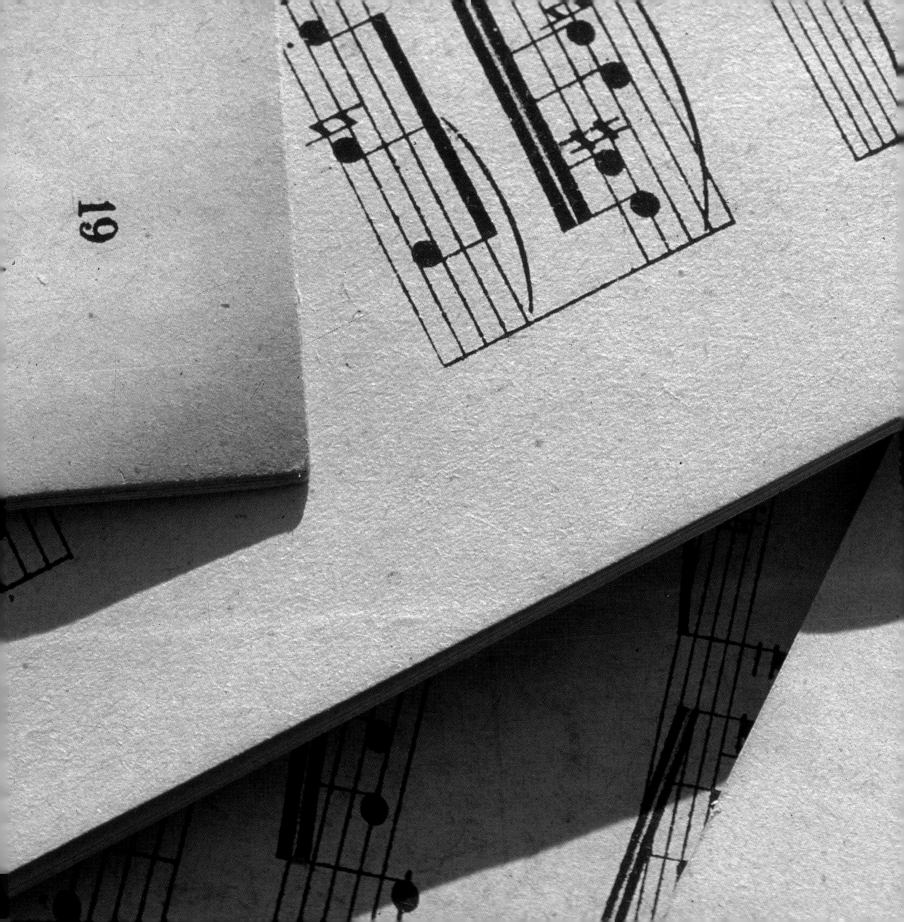

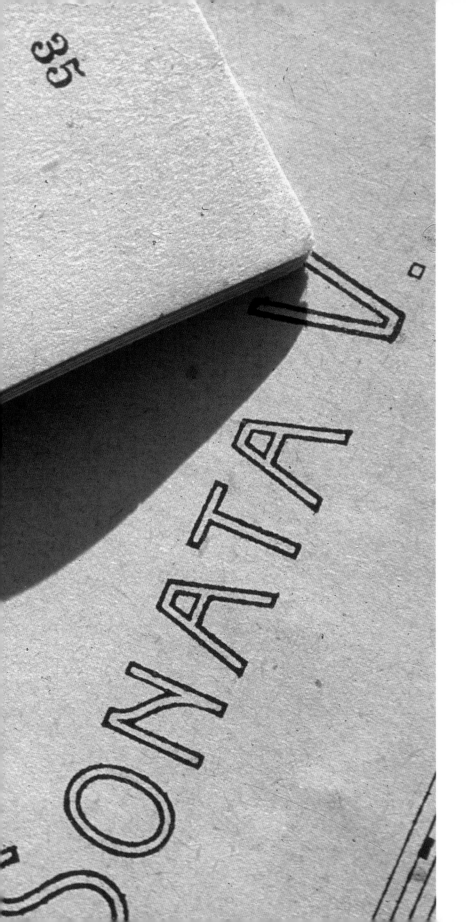

We are a clever species: we can reduce
the most complex of our joys and
sorrows to blotches of carbon patterned
on the bleached tissue of dead trees.

The process is wonderful. We imagine
the sound of music. We write the patterns of the imagined sound into
patterns of black ink on white paper.
Light shining on the patterns evokes
a spray of photons from the white
surface. The lens of our eye focuses that
spray onto the retina, which talks
to the brain. The brain translates the
conversation into instructions for the
fingers. The fingers move on the keys,
and the hair cells in the ear's cochlea
dance as the strings vibrate. The brain
listens, and finally we hear the music.

Many types of symbols on surfaces—
from cuneiform to magnetic bits, from
painting to musical notation—have
immortalized our thoughts, at least for
a while. Establishing the most efficient
ways of representing information—
music, sound, payrolls, simulations
of stars—makes the manipulation of
that information more efficient.

sheet music

○ player piano roll

We used to do much of the work of the world together with animals: the ox pulled the plow and we guided; the horse ran the mail and we rode. Now, many of the animals have been retired, we have become managers, and most of the workers—the movers and flyers and diggers and welders—are machines.

We and animals and machines are different. We talk to our own species with words, and we talk to animals with whistles and petting and blows. How should we talk to machines?

We have tried many languages: this was an early written language and involved touch. The machine, a player piano, had fingers: we wrote instructions to it with holes in a roll of paper into which these fingers fit. A hole corresponded to a note; the position on the roll was the time to play the note. This language was like Braille, patterns of dots to code meaning.

The language of the player piano was cumbersome; it was slow, and had a limited grammar. We must evolve simple, clear, and fast languages to use in talking with machines if they are to be skillful workers. The simplest current alphabet for these languages has only the two "letters" of the binary code—0 and 1—the alphabet used by all computers. The grammar and syntax of future languages will depend on the machines we address.

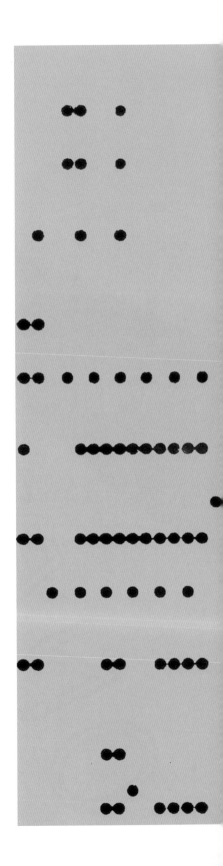

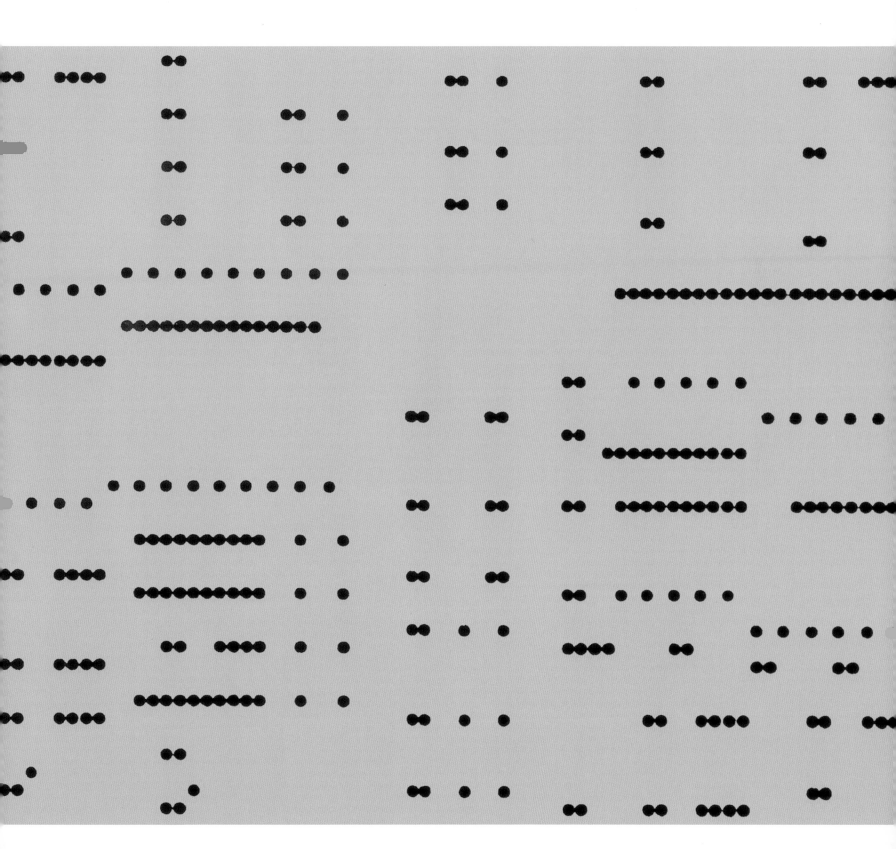

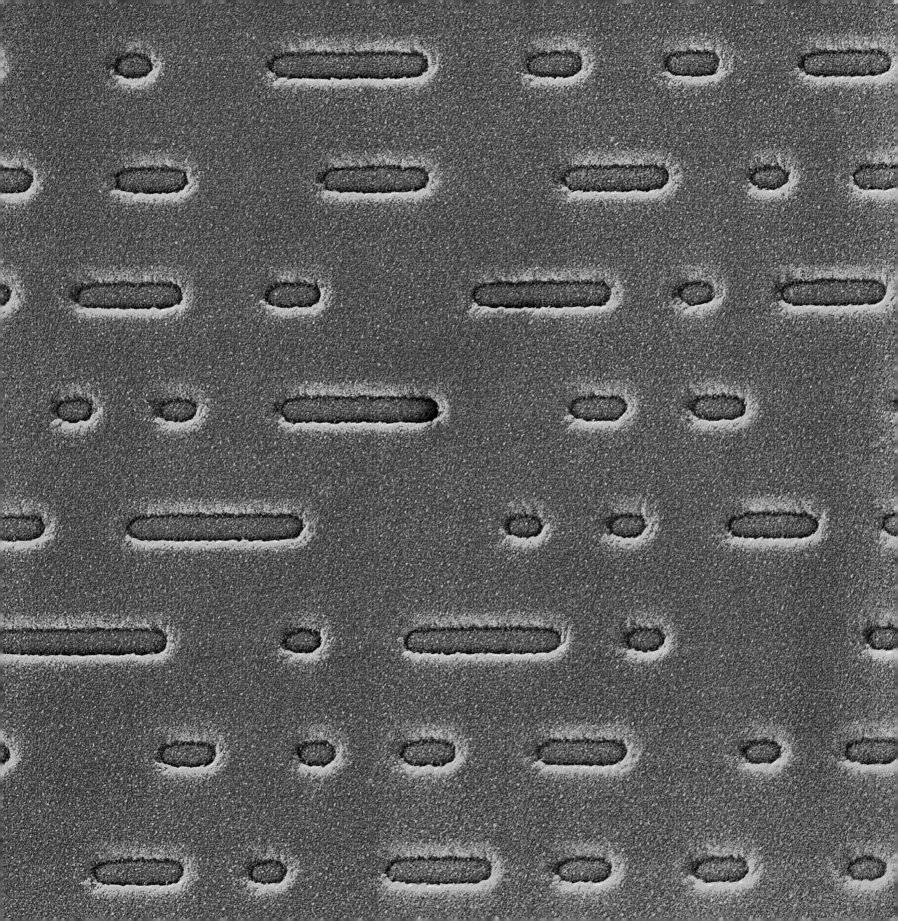

These pits are Braille for light.

The simplest alphabet is "binary code"—dots and spaces (the absence of dots), which can serve for all the letters and numbers—1 or 0, "yes" or "no." The binary code's combinations of 1 and 0 can spell out thought as clearly as the most elegant calligraphy.

The combination of eye and brain with which we read our world is a most remarkable system, capable of astonishing feats of recognition. Faces, moods, coming changes in the weather: a glimpse recognizes all. By comparison, a CD player is a one-eyed robot working in the dark with a head lamp: a simple red light illuminates the spinning disc; a simple, unblinking eye looks for reflected red light. If there is no pit, the light reflects; if there is one, it does not. Any information that can be coded by letters or numbers can be translated into the 1 and 0 of the binary code. Each letter and number is represented by a sequence of bits: the number "5" is "101" in binary. *The Well-Tempered Clavier* becomes a numerical record of the amplitude of the pressure exerted by the sound waves during performance on the recording microphone.

The challenge in a simple job is to do it fast, without making mistakes. CD players succeed. A beam illuminates the disc, and a photoreceptor registers the information faster than any human eye can appreciate, but there is no appreciation of style: just 1001011011001. *The Well-Tempered Clavier* and Led Zeppelin are both ruby glitters from a spinning disc. The light brushes the pits with the gentlest of fingers, and with no physical contact there is no wear.

Our appetite for stored information is voracious and largely unconscious. By simply changing from red to blue light to read the disc, and by decreasing the spacing between the pits accordingly, it is possible to store four times more information per disc.

compact disc

The diamond in the wedding ring is carbon, crystallized in a world of heat and pressure so extreme that we can not imagine it. Were we to be placed there—the better to observe and understand—we too would become diamond.

Diamond is now manufactured as well as mined, and the diamond mines are no longer the sources of unique treasure that they were. The newest process for making diamond is startlingly different from nature's method. The carbon starts as gas—the same used to cook hamburgers. A hail of electrons breaks the gas into hydrogen atoms and carbon-containing fragments, some of which condense with one another and with the surface. This process coats the surface with tar, but even as the tar forms, parts of it are burned away by the atoms of hydrogen.

Formation and destruction compete, and at the end, only the substance that best resists attack by the hydrogen atoms—diamond—remains.

Diamond is the hardest substance, and the best conductor of heat. It also has no affinity for electrons: if added to diamond, an electron slips off effortlessly if beckoned by any positive charge. The grip of diamond on electrons is the tenuous grip of grass on fog.

These patterns are grids of small holes in silicon that have been filled with diamond powder. Electrons are pumped from the silicon onto the diamond; once there, they stream off smoothly toward a positively charged electrode. The arrays of holes are rectangular shower heads spraying electrons. These sprays can be used to generate light, and they may be used in new types of display systems bright enough to be seen in full daylight.

diamond electron emitters

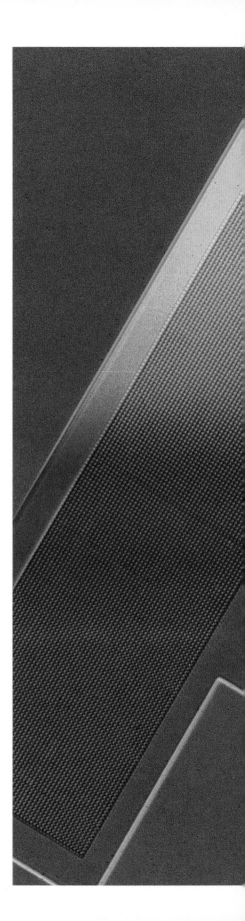

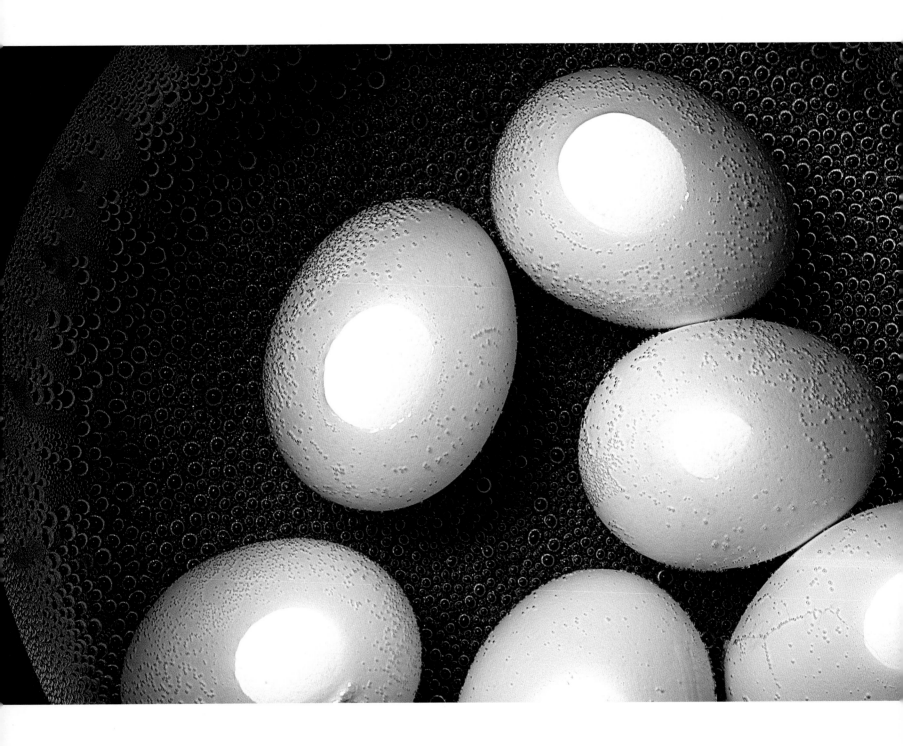

eggshells,
and breathing

Throughout our life — from conception to death — we must breathe. To live — to get through each minute of our lives — we require glucose and oxygen. In our cells, in a complicated set of reactions, the two "burn," and the energy of this "combustion" powers the cell. Viewed from a substantial intellectual distance, we are not so different from a lawn mower.

In the days immediately after conception, when we are just a few cells, we can pick up enough oxygen from the environment to live. But from our birth to our death, we use lungs and heart and blood to move oxygen from the air to the cells. Between conception and birth, we have a problem — we have no air to breathe. When we develop lungs they are filled with amniotic fluid, and we rapidly become too big to pick up oxygen by scavenging from the environment. Our solu-

tion to this problem is the placenta, which transfers oxygen from our mother to us; she breathes for us. As adults, the membrane across which oxygen passes is in our lungs, inside our chest; *in utero*, this geometry is reversed — the placental membrane is outside our body.

What about a chicken? As an unhatched chick, it also has no air to breathe. It has no cooperative mother or nourishing placental membrane — it is on its own, defenseless in an unwelcoming world. Its yolk and white provide the food it needs. Its shell provides mechanical protection — without it, the chick would be overwhelmed by insects, bacteria, and

mice. But what about oxygen? The shell again does the trick. The shell looks hard and impervious, but it is, in fact, porous — a network of tiny channels connects the inside of the shell to the outside. One function that it serves is to let air — particularly oxygen — cross through these tiny channels to the inside. That oxygen is picked up by blood vessels in the chorioallantoic membrane, a structure just inside the shell, and carried to the developing chick.

These unfertilized eggs (without embryonic chickens inside) are being cooked. As the water heats, air in the shells begins to expand and escape. As it does, it forms microbubbles, and these microbubbles nucleate the growth of larger bubbles on the surface of the shell.

analysis of DNA

When game animals are driven through the brush by hunters, they emerge by size. First come the small ones — weasels, rabbits, foxes — that scuttle and slip and weave among the branches. Then come those of medium size — pigs, goats, deer — that pick their paths more carefully, trying here and there. Finally come the large animals — elk, bear — that must force their passage.

These patterns show molecules of DNA that have been driven through molecular brush: a thicket submerged in water. The process is perhaps more like driving eels than four-legged animals, but the principle is the same: the DNA is the quarry, moving by size.

The thicket is a gel, a loose mesh in which the strands are long, string-like molecules knotted into a three-dimensional net and suspended throughout a pool of water. The molecules making up the mesh are chosen

to give a net with holes of the right size to obstruct large molecules of DNA, but otherwise not to interfere with it. All the DNA is initially placed at one end of the pool. The molecules of DNA have a negative charge. When a voltage is applied to electrodes placed at the ends of the pool, they swim away from the negatively charged electrode and toward the positive. The smallest molecules do not even notice the mesh: they simply stream straight ahead. Those of medium size slow down: they wrap around occasional strands briefly, but soon they undulate free and move on. The largest molecules entangle badly and move ahead only slowly.

Each band shown here is a collection of DNA molecules of the same size. The separation of molecules of DNA by size is the basis for almost all procedures for genetic analysis, used for finding the fathers of children of uncertain paternity; for identifying the dead; for ferreting out criminals; for warning of genetic damage; for typing cancer and detecting AIDS; or for showing that the scallops on the dinner plate were born shark.

82

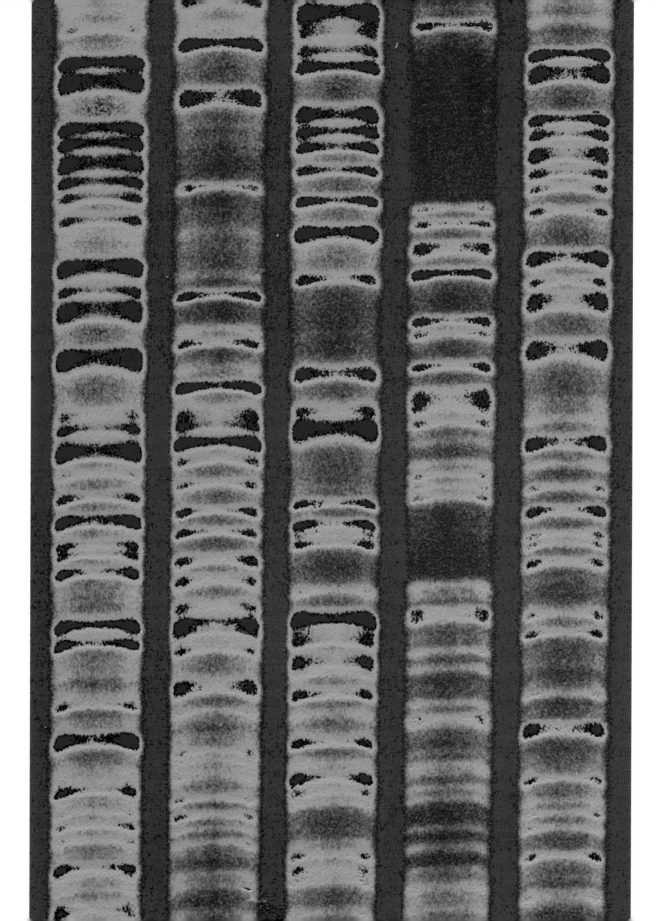

83

change

We know "now";
we remember "then."
If the two are different,
that is change.

We are immersed in a world of clocks. All that happens meters
time. The universe expands, continents collide, we age,
leaves change color, the sun sets, children play, we breathe, the
wind rustles in the grass.

We know "now"; we remember "then." If the two are
different, that is change. Without movement of the clocks—
if we know *only* "now" or *only* "then"—time freezes.

Sometimes a thing relates its own history. The trail of a
meteor, the bands of an agate, the rusted tool—all tell, at the
same time, how it is and how it was.

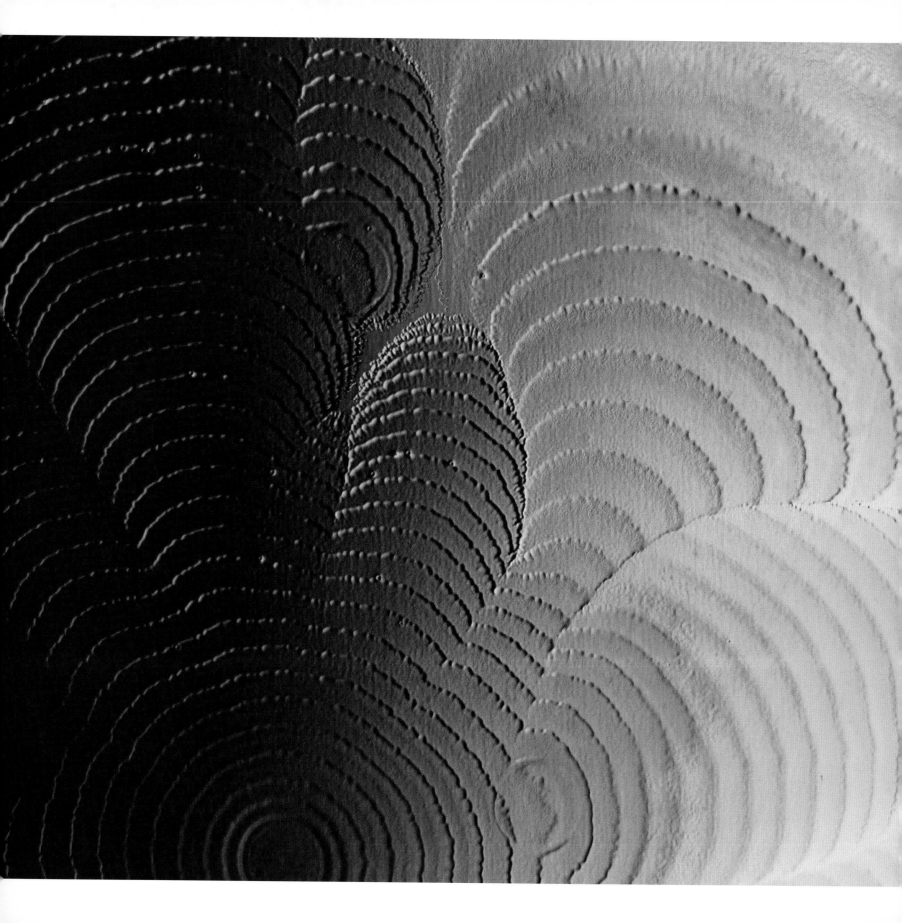

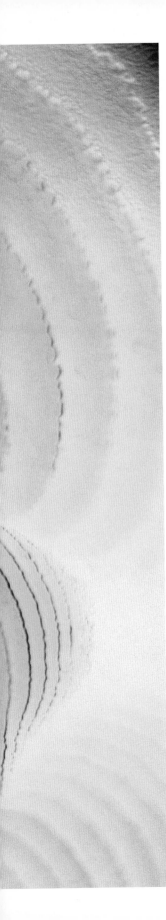

This dish records the history of a civilization of bacteria—the few days that it lasted. The civilization started at the center, when a fresh plate of culture medium—a gel containing nutrients—was inoculated with a few bacteria. They were the first colonists. The population of the initial colony grew, and it depleted the food that was available locally. When the population became large enough, new colonists were sent out. These bacteria flowed radially outward toward more abundant food and uncontaminated space. Following their wave of migration, the colony consolidated its growth in the new region until it reached the critical density in population required to send out colonists again. These cycles of expansion and stationary growth created the patterns here.

The sectors follow the history of particular subgroups of these bacteria. Some families grew to the edge of the plate; less successful families were overwhelmed by the more successful, and they faded. The bacteria could not talk with one another, but they interacted by the chemicals—the "smells"—that they left, and by their influence on the local food supply and the level of pollution. When the food on their continent was exhausted, they died.

From these patterns, we infer how bacteria sense their environment. We sample *our* environment with different and more highly evolved senses than they do, but there are remarkable similarities between their sensors and ours. Understanding perception in all its complexity is one of the great challenges now facing biology. Evolution connects our eyes and nose to the senses of bacteria; their present is our history.

migrating bacteria

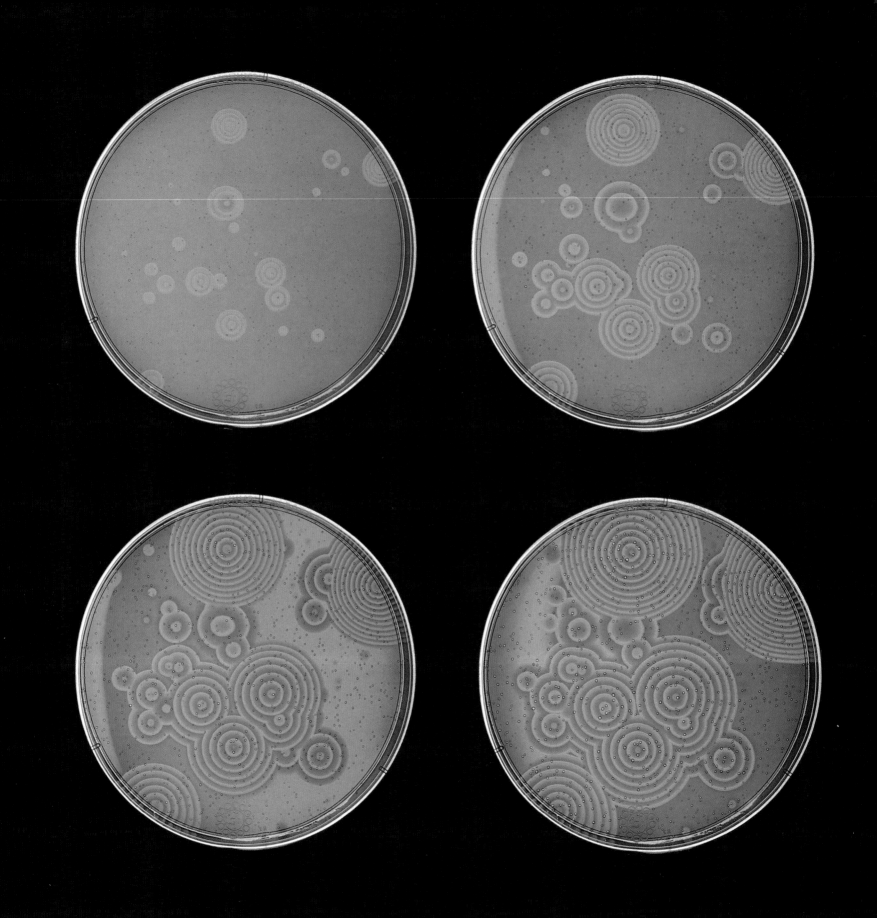

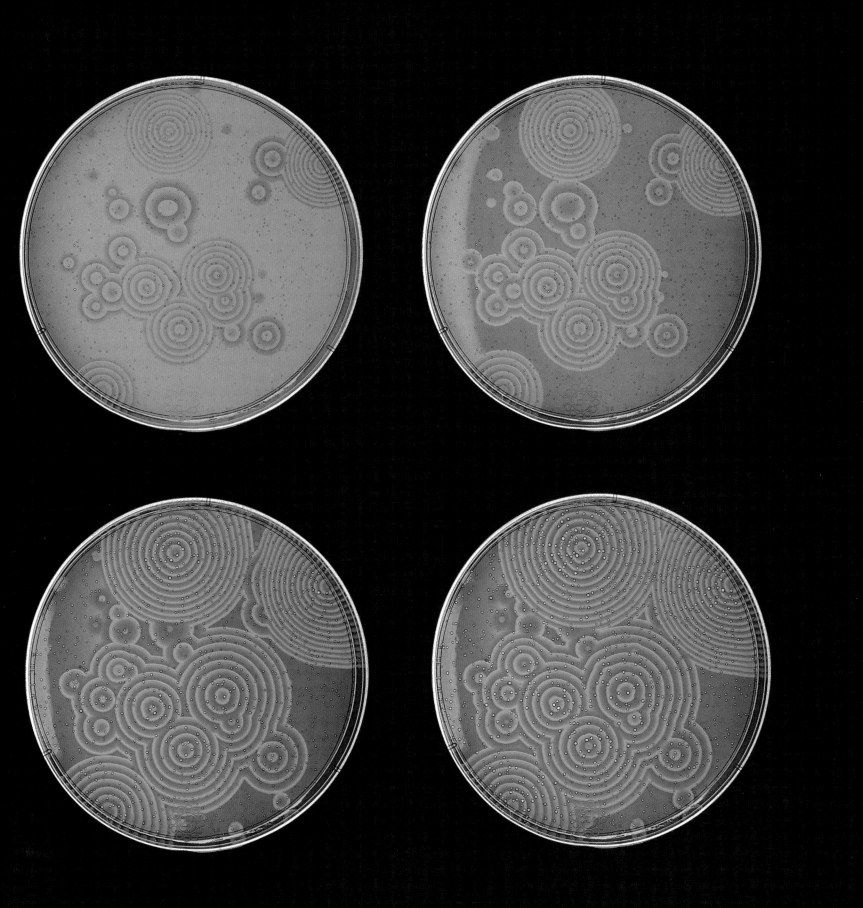

Waves of chemical reaction ripple outward on the surface of a chemical pond. To a stationary observer staring at a point on the surface, the colors—mirroring the concentrations of reactants and products—oscillate. These patterns are startling in their regularity and complexity.

A solution of the reactants was poured into a dish. Reaction did not occur uniformly; it initiated at stationary spots—perhaps on specks of dust, perhaps by accident—and spread as concentric waves. The first product formed in an autocatalytic reaction: the first small amounts of this product that formed accelerated the formation of more of it. The first reaction swept outward, generating high concentrations of the initial product as it moved. A second reaction then occurred that involved the product of the first as a reactant. This second reaction followed the first, destroying the initial product and forming a second. The combination of these two processes—

a wave of the first reaction forming an initial product, and a subsequent wave of the second destroying it— caused successive waves of reaction to ripple outward. An indicator molecule—an observer of these processes— signaled the passage of successive waves of reactions, turning from red to gray as the first reaction took place; and from gray back to red for the second. These images of the same disk show the evolution of these reactions.

The mathematical description of these reactions is analogous to those used to describe predator-prey relations involving foxes and lemmings. First, the lemmings eat grass, and their number grows. The foxes find the abundant lemmings to be a fine supply of food. The population of foxes grows and that of lemmings declines, until there are no longer enough lemmings to support the foxes. The population of foxes then declines, that of lemmings again increases, and the cycle repeats itself.

Oscillation is associated with control, although control is not an important part of the processes occurring here. Steering a bicycle while weaving it back and forth across a line is easy; keeping it exactly *on* a line is difficult. Electronic circuits, chemical reactions, and bacterial metabolism are often most stable when oscillating around an average value—a control point.

These stable oscillating patterns catch the eye (a black and white zebra attracts more attention than a gray pony), and demonstrate how chemical energy can be converted to unexpectedly regular patterns, and how oscillations in time can be converted into oscillations in space. They hint at structure in life—the most spectacular of processes that burn chemical energy and generate complexity. Some of the reactions occurring in simple organisms seem to oscillate, and more complex regularities exist throughout nature: the pacemaker of the heart, the trains of spikes nerve cells use to talk to the brain, the aggregation of slime molds, the patterns in the fur of animals, and the extraordinary geometrical organization that characterizes the early stages in development of an embryo.

oscillating chemical reaction

PREVIOUS PAGES

90

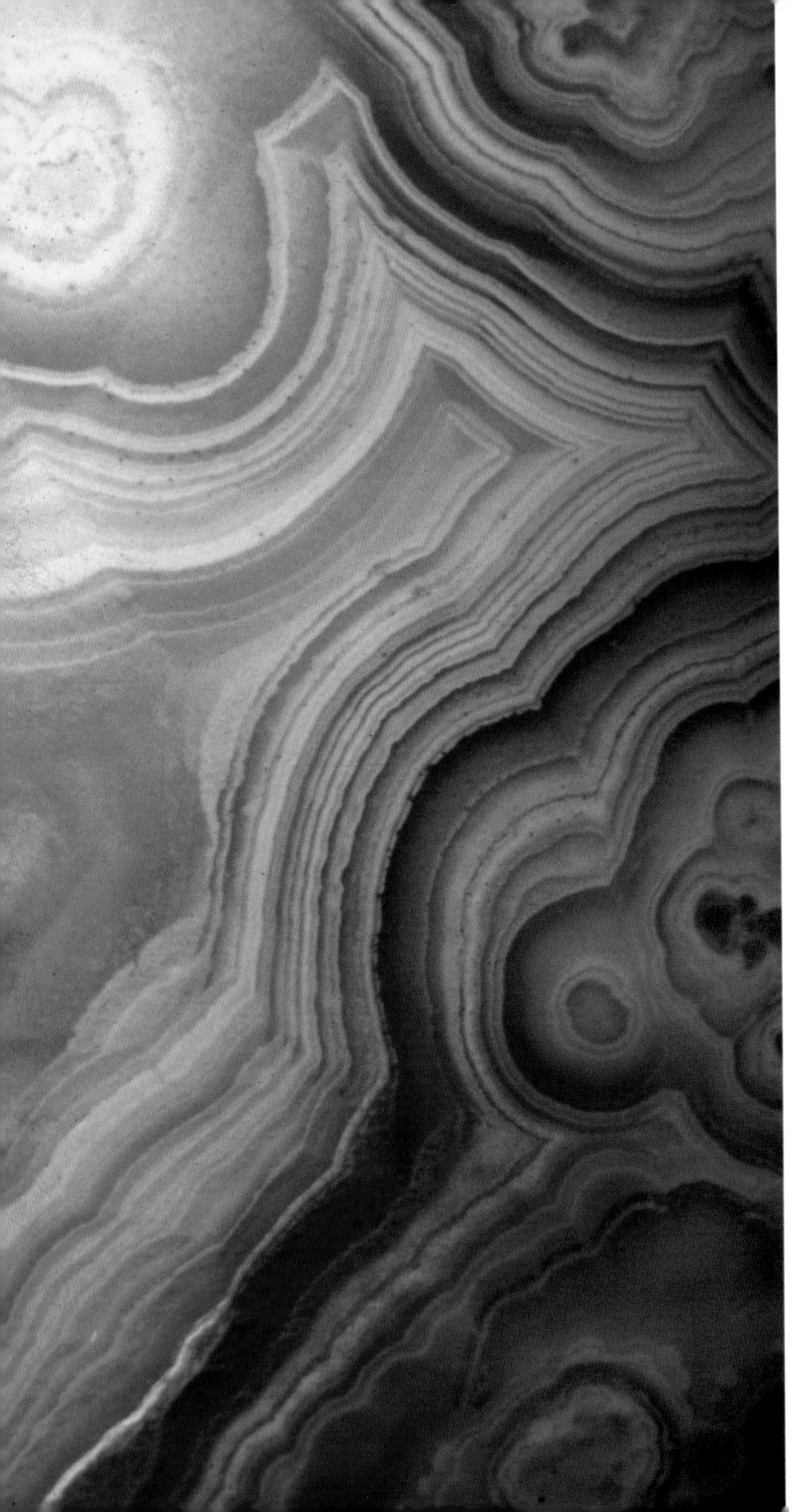

agate ○

The subtly colored bands of agate and
the layers of accumulated muck that
plug the sewer pipe of a house both
record sludge formation. The pattern
shown here began with a hole in a rock.
The hole filled with water carrying dissolved silica—the substance of beautifully transparent quartz—but here it
was contaminated with other dissolved
minerals. As the water evaporated or
cooled, a layer of solid silica coated the
walls of the hole: repetitions of this
process over hundreds of years produced the cloudy record we now so
admire. The color and thickness of the
bands are a history of the composition
and temperature of the soup from
which they came, and a record of the
geological past.

mammalian cells
on a patterned surface

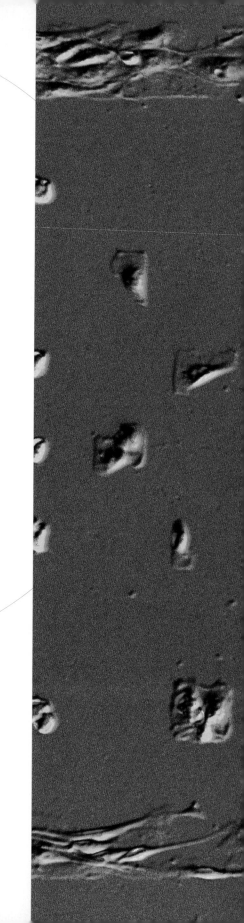

We avoid death. We think of it as an undesirable accident in the plumbing: an artery clogs or bursts; a tumor pushes the machinery out of line; an automobile at the wrong time and place reduces us to mush. Still, proper maintenance forestalls the inevitable, and we are emotionally disposed to live forever.

Many of our cells take the opposite point of view. They are always poised to die, and only constant encouragement from their neighbors keeps them living. Unless stimulated by chemical messages—growth factors—and the reassuring touch of their neighbors, many cells quietly commit suicide. We believe, as a sentient collection of cells, that our normal option is "Live, unless misfortune befalls." For most of our cells, the normal option is "Die, unless instructed otherwise." Sentience denies the inevitable.

Cellular suicide—apoptosis, or programmed cell death—is one of the profound discoveries of molecular biology. Many events can trigger apoptosis in a cell: damage by a virus or radiation, mistakes during replica-

tion, obsolescence, or an abnormal environment. Cancer is, in part, a failure in apoptosis: a cell that should recognize serious dysfunction and commit suicide fails in its duty and multiplies instead.

One of the many questions that a cell asks about its environment to determine whether all is right is, "Am I attached to something?" The normal state of most cells is to be attached to neighboring cells or biological structures: cells that wander are often deranged. The cells pictured here rest on a surface that was patterned by printing; the "inks" used permit the cells to attach and spread only in certain areas. The smallest squares force cells into entirely unnatural shapes: too small, straight edges, sharp corners, no other cells in physical contact. In this uncomfortable closet some cells continue to live but refuse to divide; some commit suicide. Observing these responses clarifies how a cell sees its world and how it chooses among its options. Understanding how cells choose to live or die will, we hope, someday help to control cancer, forestall aging, and regenerate complex organs.

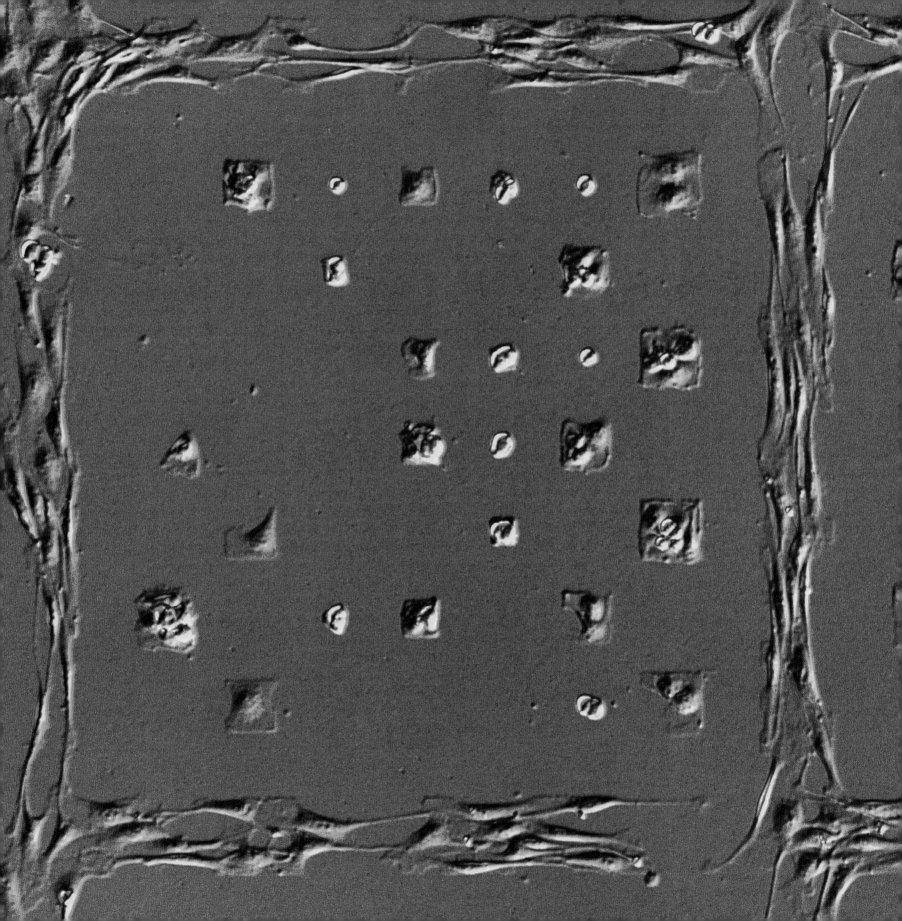

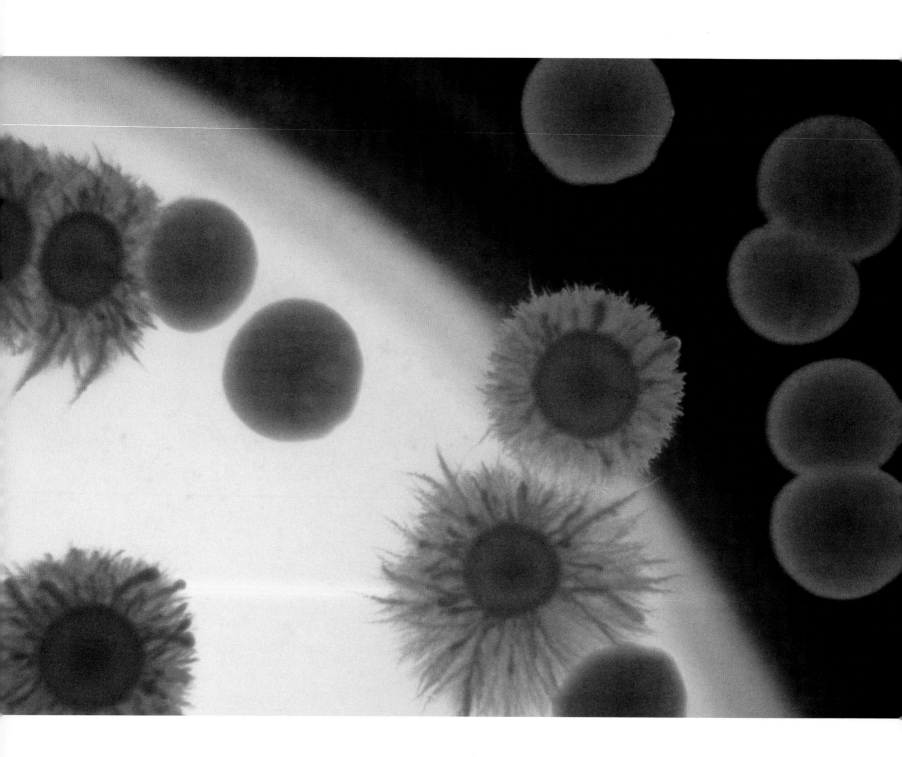

Wars start at borders. A state, once peaceful but now driven to aggression by motivations and provocations that are often obscure, crosses a border and invades the territory of a neighbor.

The relations between a disease-causing microorganism and one of us can follow the same course. These white plaques are colonies of a yeast, *Candida albicans.* This yeast can cause disease—mild in otherwise healthy people, serious in individuals with weakened immune systems due to cancer or AIDS. This yeast exists in two forms. One is smooth, one filamentous; in the laboratory, the smooth type forms rounded colonies on top of the medium used to grow it; the filamented form is invasive and launches spears of connected cells—raiding parties—into the medium. This transition from the smooth form to the filamentous, invasive form is associated with disease in humans.

What provokes the yeast to aggression? The signals that cause a smooth yeast to convert to an invasive form are chemicals in its environment; these chemicals are sensed by the yeast using something like a sense of smell. Detection results in a cascade of internal changes that cause the transformation. Its tendency toward aggressiveness is hereditary: the molecules that sense the chemical signals are coded in its genes. If these genes are damaged, the yeast no longer responds: it remains the smooth, noninvasive form even in an environment that would provoke a normal yeast to invade.

In this image, the normal colonies of yeast cells have transformed to filamentous growth, and the abnormal colonies having defective genes have remained smooth, unable to transform. Understanding the world of the yeast may suggest strategies for defeating the diseases caused by its invasions.

colonies of yeast

Swirl wine in a glass and watch: a film of liquid creeps up the wall of the glass, pauses, collects itself into drops along the top of the film, and slides back down into the wine. The motion evokes the fluidity of whales breaching or the changing shape of fog condensing into rain. It is an astonishing amount of independent activity for a simple liquid: a kind of levitation followed by lateral condensation. It smacks of water running uphill.

The invisible hand that pulls the wine up the side of the glass is surface tension: the tendency of a liquid surface to contract and to minimize the number of its molecules that are at a surface exposed to air. On the wall of the wine glass there is a competition. The solid surface of the glass tries to minimize its exposed surface by covering itself with wine; the pull of the glass on the wine generates a thin liquid film and increases the exposed surface of the liquid. The wine meanwhile tries to shrink its surface.

The competition between stretching and shrinking is complicated by the fact that the wine's eagerness to contract is different at different points. Wine is a mixture of alcohol and water. Water by itself does not tolerate being spread as a thin film — it has a very high surface tension and strongly resists having its surface stretched — but the alcohol in the wine weakens this resistance. The molecules of alcohol at the surface of the liquid attract their neighbors only weakly, so stretching a film of water containing ethanol is easier than stretching pure water.

The alcohol is also volatile: it evaporates from the wine and escapes into the air. This evaporation is most rapid at the lip of the film of liquid that covers the wall of the wine glass, since this lip is closest to the open top of the glass. When the alcohol evaporates, the proportion of water in the remaining wine increases, and its surface tension increases. This wine enriched in water collects itself into drops, pulled together by its increased surface tension. These drops slide back down the wall of the glass into the wine.

Phenomena related to tears of wine are important in many technologies that involve thin fluid films, especially if these films have components with different volatilities, or are exposed to different temperatures, as in distillation, or the spreading of paints containing volatile solvents.

tears of wine

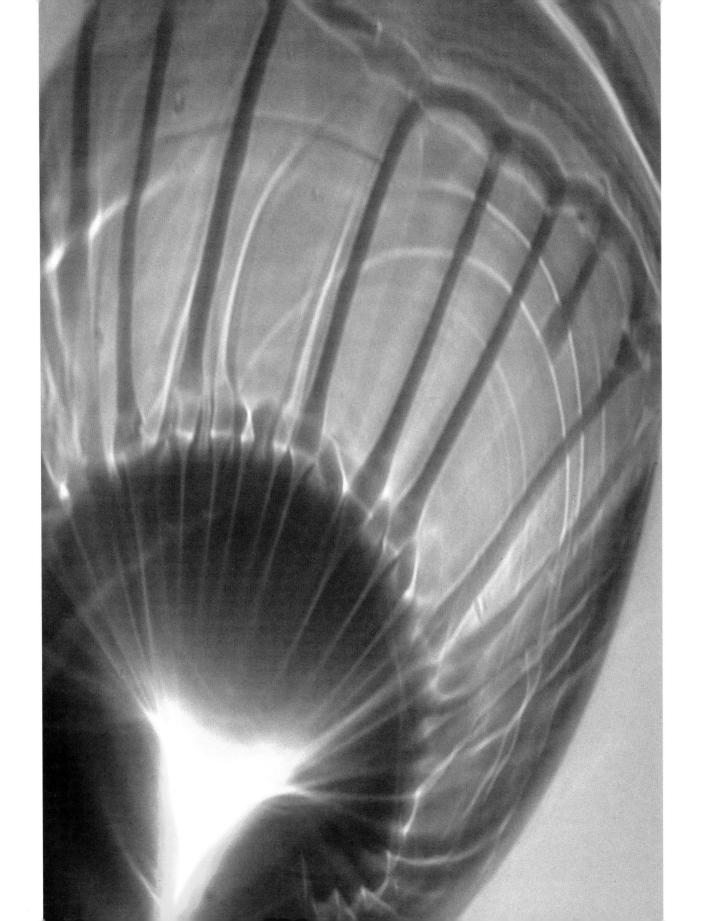

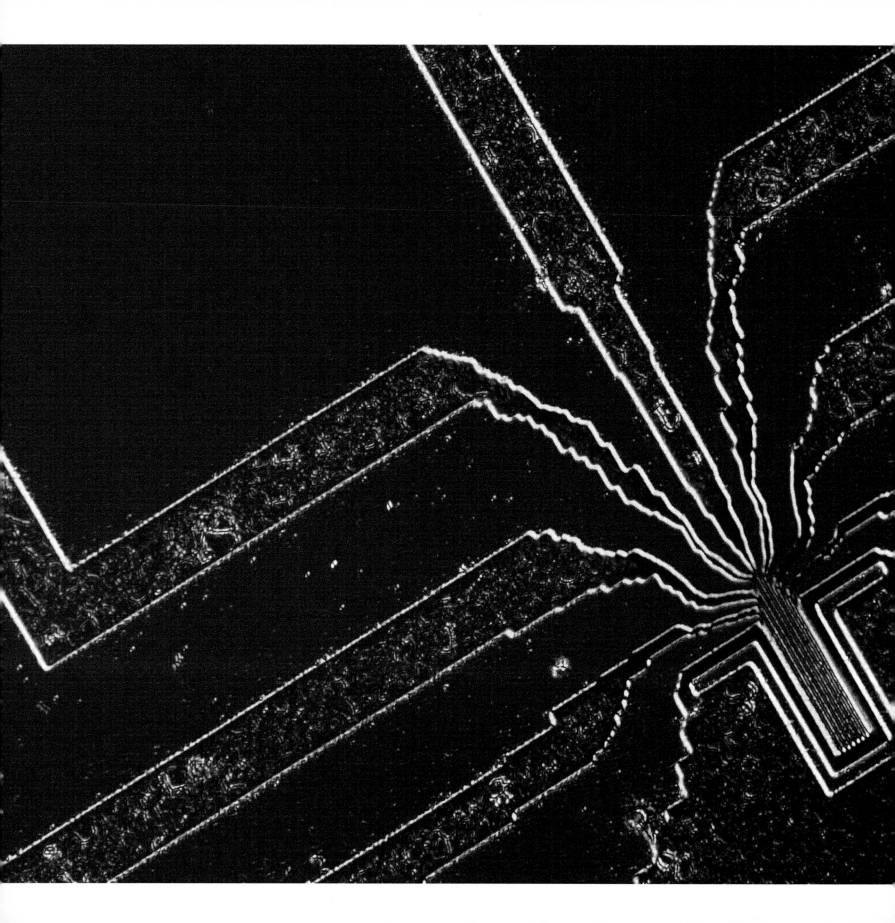

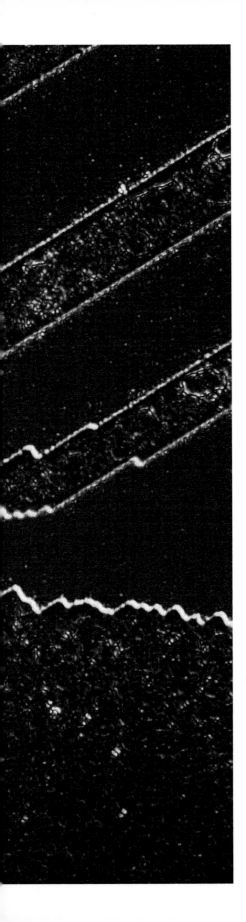

microelectrodes

Electrons know two verbs: *seek* and *avoid*. They seek the positive charges of atomic nuclei; they avoid the negative charges of other electrons. That is almost all they know.

To electrons, a solid metal wire is a hollow pipe through which they can flow. Here, a pattern of small, unconnected wires was immersed in a liquid: in the image, the wires are the flat brown bands outlined in gold. The smallest wires—those with which the experiments were done—are barely visible in the "stem" on the lower right. Electrons were forced into the ends of some of the pipes and sucked from the ends of others. If the pressure on the electrons was sufficiently high in the pipes that were overfilled, they squirted onto molecules in the surrounding liquid.

The molecules carrying these escapees moved randomly through the solution. When they approached one of the partly emptied pipes in their meander-

ings, the electrons sensed a vacuum— a deficiency of electrons—and hopped in. It was a kind of children's game for the electrons: crowd into one pipe; hop out onto a molecule in the solution; wander with your molecule; find an "empty" pipe; hop back in; escape.

This lively process is useful for studying the molecules in solution, not the electrons themselves. The movement of the electrons in the wires—the electrical current—is well understood. The amount of pressure in the wire "pipe"—the electrical potential— required to force them out of the pipe and onto the molecules in solution is determined by how welcoming those molecules are and is one way of characterizing them. The time required for the molecules with their electron passengers to drift from one electrode to the second tells how rapidly they move. This information helps those who design batteries and those who use electricity to cause chemical reactions. It may lead to improved sources of portable power for running laptop computers and cellular phones for longer times and for silent electric vehicles contributing little to urban air pollution.

99

When we die, we leave behind a midden: photographs, bank accounts, letters, clothes, teeth, bones. Whatever patterns the artifacts in these piles of rubbish carry are usually and mercifully lost as they are mixed into the compost heap of time past. Biological artifacts are especially evanescent: as fire eats wood by oxidation, so air eats paper. Librarians even call this process "slow fire."

Shakespeare's signature

We occasionally wish to preserve artifacts that remind us of the famous or notorious. Conserving antiquities—paper, tissue, fabric—is a part of recording history and of keeping our past from being eaten by ghosts. Sophisticated technology now preserves very old papers and fabrics, and establishes their authenticity. In this instance, identification of the signature as authentic is based on a web of evidence that includes measurement of the rate of migration of iron (present in the ink) away from the edge of the line: the distance over which the iron migrates depends upon the length of time that the ink has been in contact with the paper, and helps to date the signature.

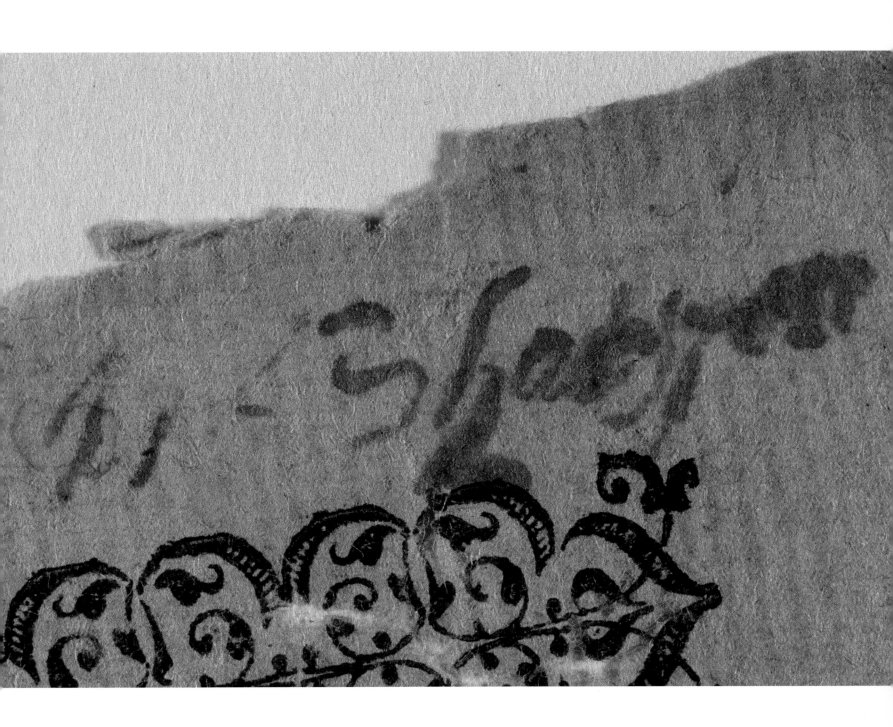

disorder

It all ends in ruin.
We design, we build, we conserve,
but the end is always disorder.

It all ends in ruin. We design, we build, we conserve, but the
end is always disorder. We extract order from disorder:
we separate, refine, shape, assemble. What we build, we protect.
But milk and coffee mix in the cup; iron and oxygen combine;
machines wear and fall apart. The final state is disordered,
not ordered; mixed, not separate. Our constructions are
the triumph of effort over the inevitable; their decay is the
triumph of the inevitable.

Mixing—the growth of disorder from order—involves
surfaces. The first signs of disorder, failure, and decay often
occur at surfaces joining unlike things, such as the man-made
and the natural. We observe the processes that lead to
disorder—failure, fracture, wear, corrosion—and learn,
sometimes, to forestall the incursions of chaos.

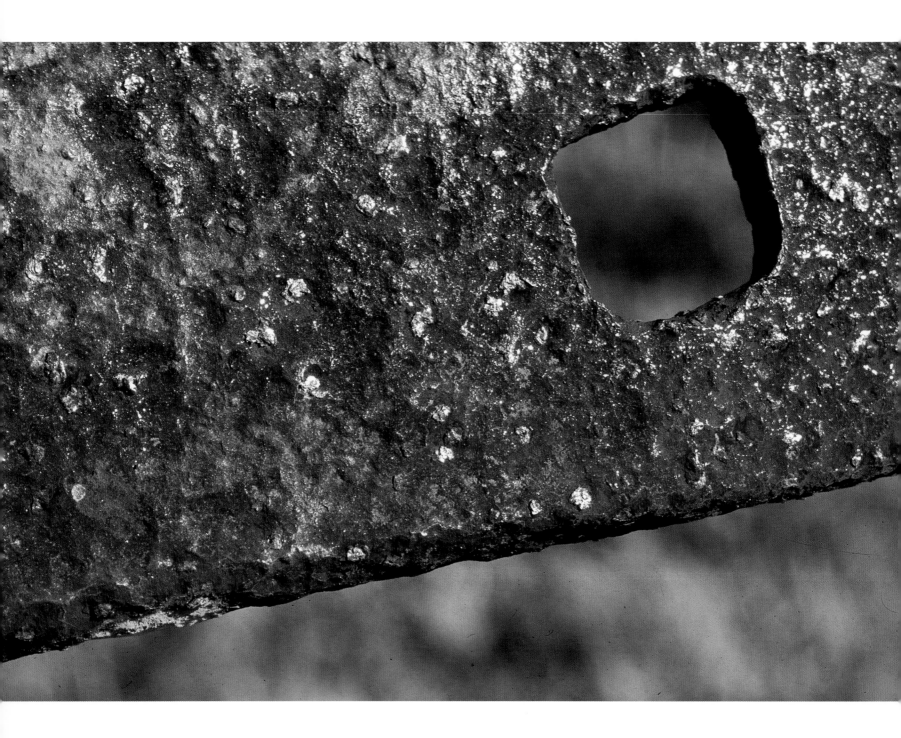

rust

To make a new piece of iron—a tool, or part of a structure—requires us to spend energy in great quantities: energy to extract the iron from the ore, energy to shape it. As quickly as we are done, the process reverses: very, very slowly, the iron burns, oxidizing where it is exposed to air and moisture.

Iron rusts; we decay. We, and iron, and stars, and the universe itself all move at our own pace to that inescapable final state of disorder. We obey few laws, but one we all obey is the second law of thermodynamics: Where we end is where disorder is the greatest.

The process occurs from the outside in: the oxygen in the air must reach the metal. The rust forms and falls from the surface in disorder; the smooth surface roughens and turns the dark red-brown color of iron oxide. Lichens graze in the ruins. The form of the piece diminishes, and finally, in a leisurely way, disappears. Rust to rust.

This piece of silicon was immersed in a chemical ice storm: a fog of reacting vapors that condensed as a tough glassy film on its surface. The film was intended to keep the herds of electrons in the silicon from escaping, and to protect them from harm. It was to have been a secure pen: a kraal for electrons.

The procedure failed. The film formed, but it shrank. Instead of sticking to the surface, it ripped loose and peeled like sunburned skin. In the rubble lies instruction. The scholars of this type of failure look at the fragments and judge the misfit between the skin and the surface.

The art of building microelectronic devices is the art of making very large numbers of microscopic things perfectly: the gates, transfer chutes, and holding pens for electrons must be small, and they must not leak. The devices operate by moving electrons from place to place, as a stockyard moves cattle. The devices are very small, and the electrons need move only short distances—a few nanometers or a few millimeters. Small distances mean short times, fast calculations, small quantities of materials, and low costs.

mechanical failure of a thin film

FOLLOWING PAGES

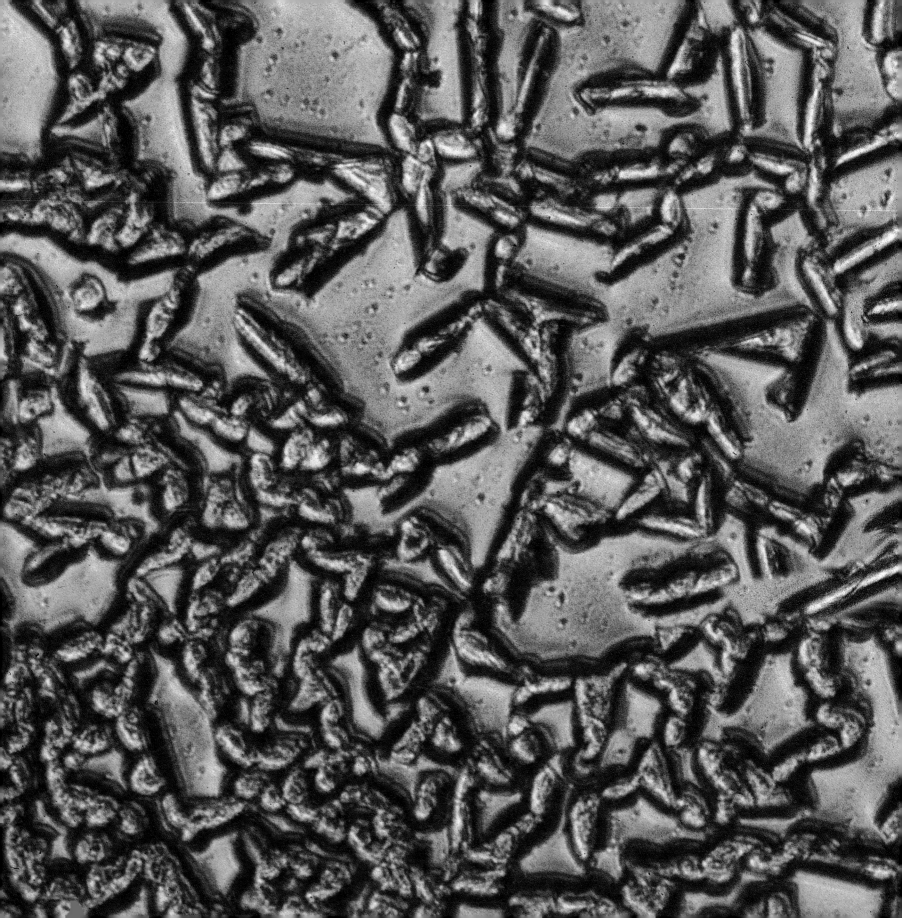

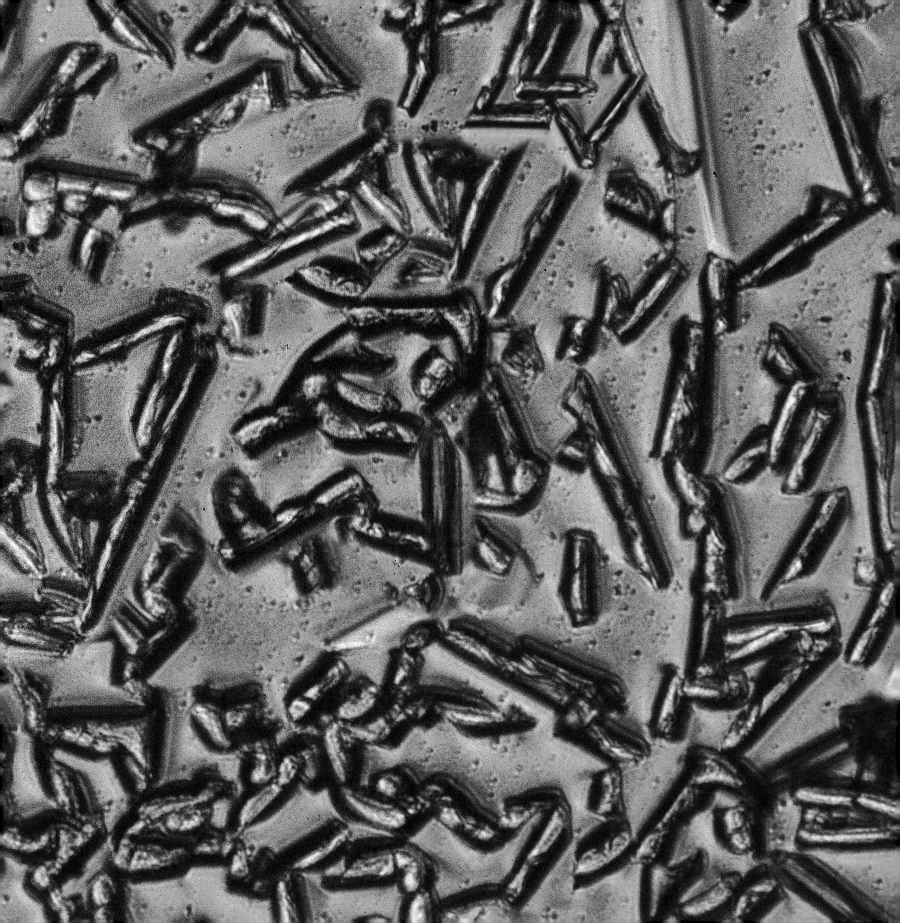

One of the small ecstasies of child-hood is to see what can be set on fire using a magnifying lens and sunlight. Even a small lens, held with a steady hand, can focus the image of the bright summer sun and set a dry leaf on fire, or cause unconscionable distress to ants and beetles. The lens creates a miniature solar furnace, and with it children can wreak the minor random acts of vandalism and cruelty so characteristic of the unsocialized early years of our species.

But children playing with children's toys usually do damage only by accident (a burning leaf *can* cause a burning house). What can be done intentionally with serious adult tools? These spots were created with short pulses of very intense light from a laser focused *inside* a slab of transparent silica. Normally light simply passes through silica—the stuff of which quartz is made—but when the light is sufficiently intense, strange and unexpected things begin to happen. The electromagnetic field of the light is so strong that it tears electrons from the silicon and oxygen atoms making up the silica. In so doing, the laser deposits so much energy in the

microexplosions inside silica

minuscule volume of sharpest focus that it causes almost unimaginably rapid local heating. The details of what happens in these experiments are not entirely clear, but in about 100 femtoseconds (less than a millionth of a millionth of a second), the silica heats from room temperature to a temperature at which it explodes: a tiny sun is born and dies at the focal point of the light.

These spots (which are smaller than can be imaged accurately with a microscope) record a test of the ability of this type of explosive local heating to produce a regular grid of cavities inside the silica. They are generated one at a time by moving a silica slab under the laser as it fires. These arrays of cavities, patterned in three dimensions, might be used to store information for very long periods of time, or perhaps to make new kinds of optical devices.

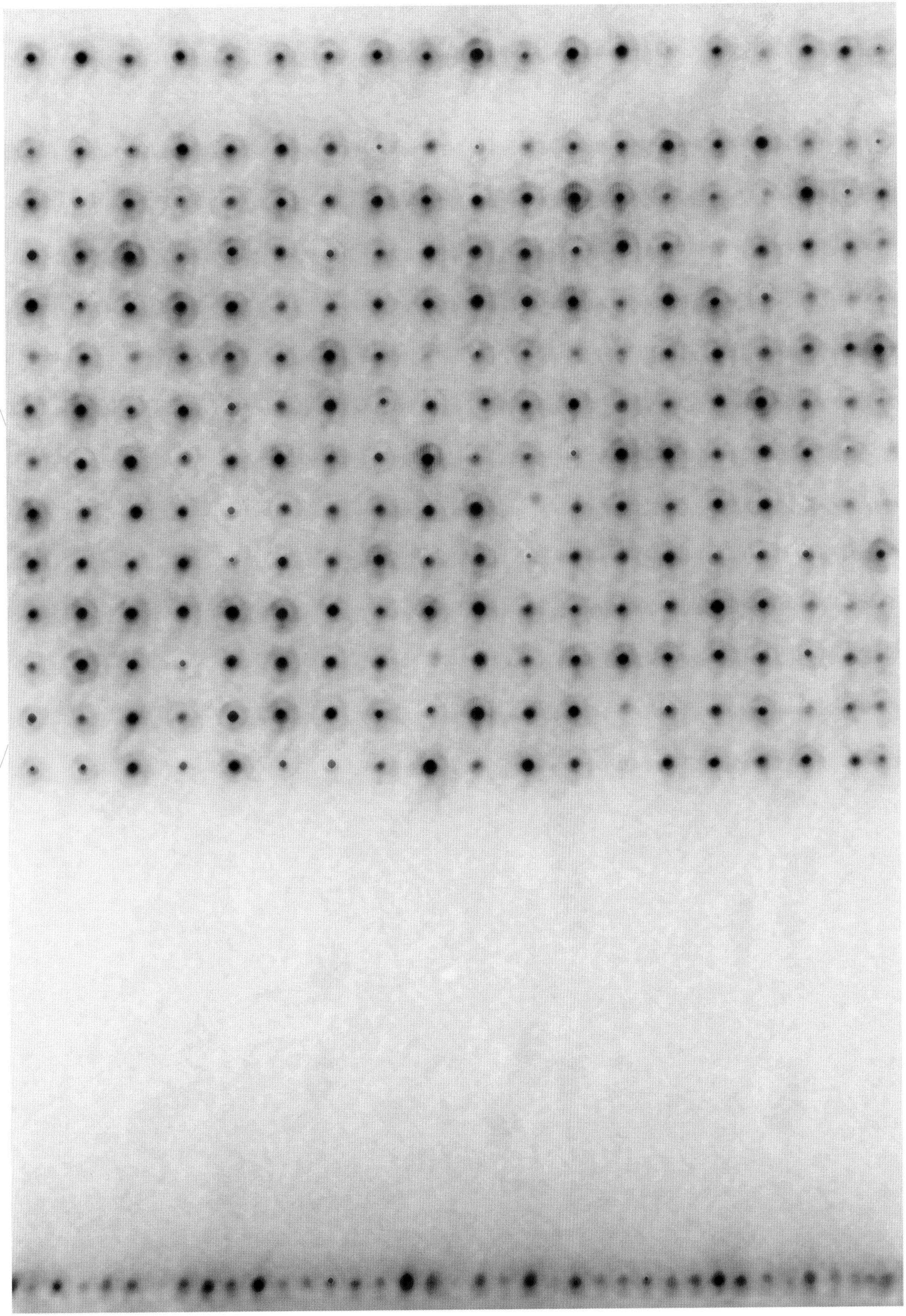

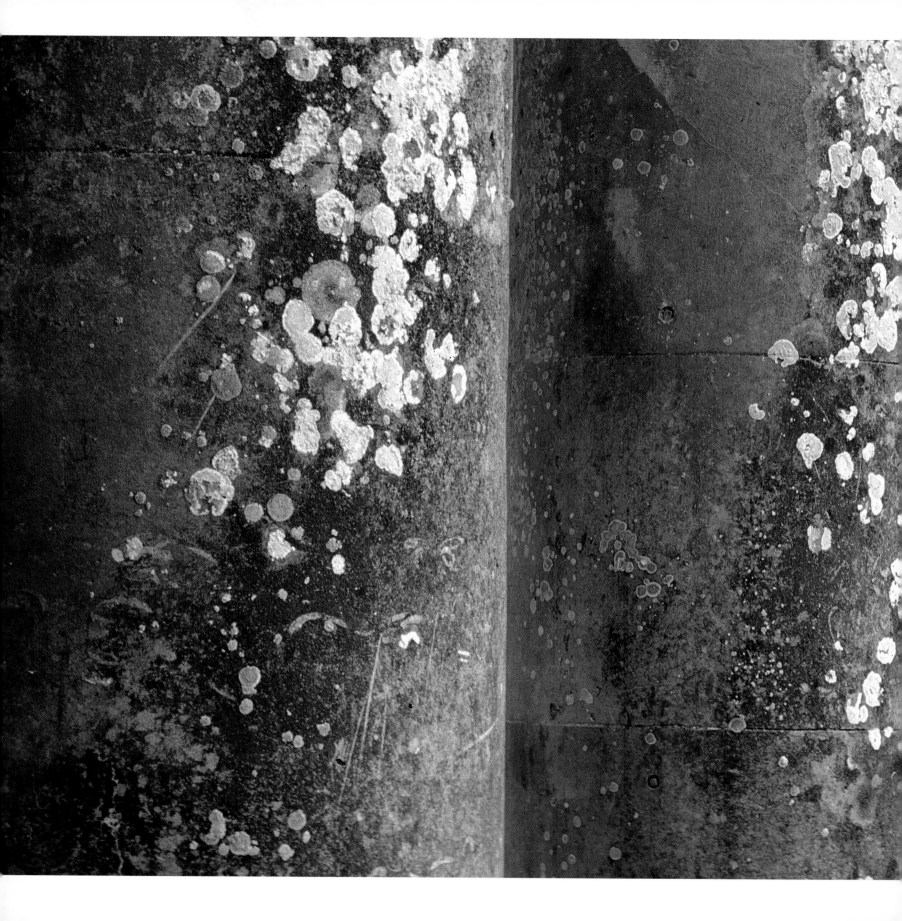

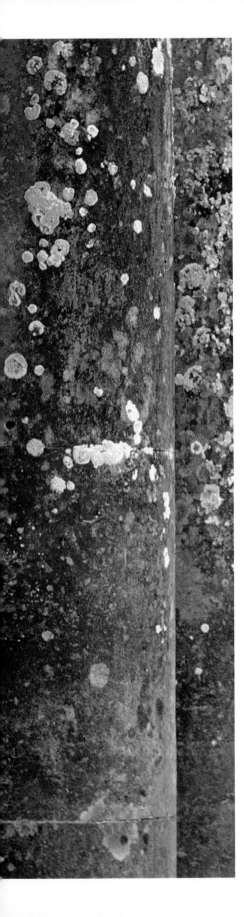

Ancient eaters of rock; microorganisms
that patiently digest the most unpromis-
ing of foodstuffs and convert it
into the most unprepossessing of life
forms. To us, the stone columns
decorate an elegant formal garden and
signify the ennobling influence of
culture and the enrichment of the spirit;
to them, the columns are lunch.

All living things must have minerals—
phosphate, sodium, potassium,
calcium—to make their internal
machinery work. We obtain these minerals by eating plants and animals;
plants obtain them from soil; lichen
extracts them from the least cooperative
material—rock—and from rain-
water and dust.

Lichens are a cooperative venture:
they are a symbiotic mixture of two
forms of microorganisms: algae and
fungi. The algae harvest light photosyn-
thetically and provide the organic
materials necessary for the fungi; the
fungi produce acids that leach minerals
used by the algae from the rock.
These slow processes erode many
forms of rock and are a subdued part
of the ecology of the planet. They
eat the stones we carve.

· columns with lichen

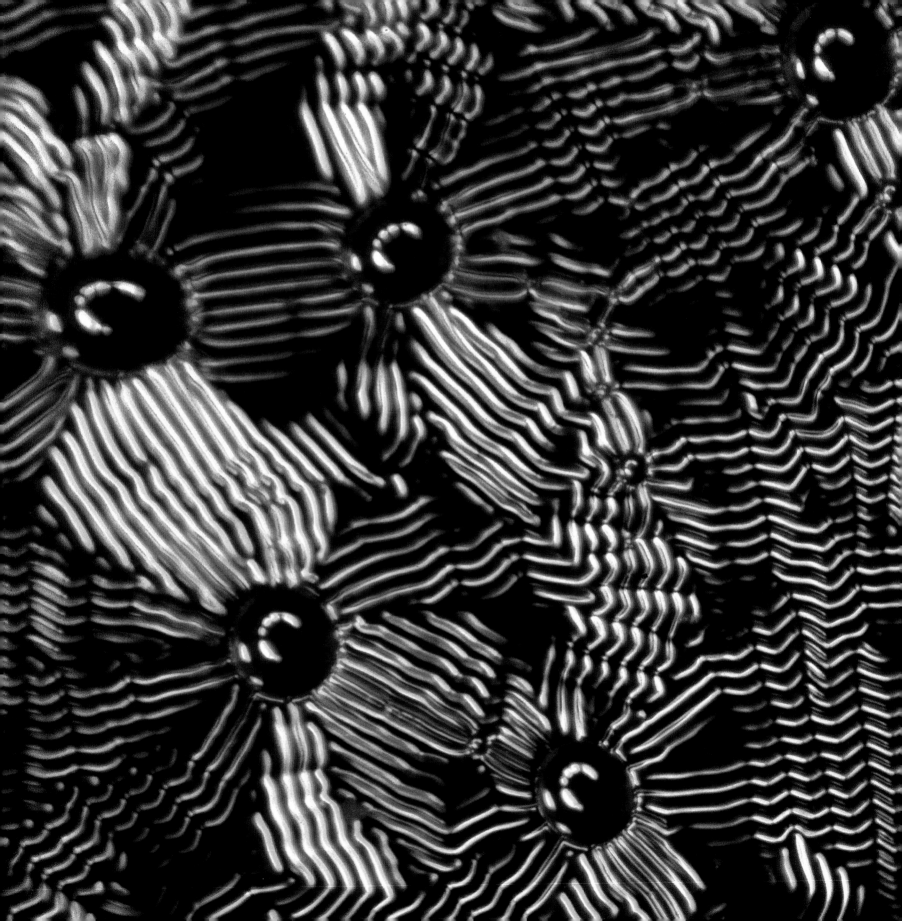

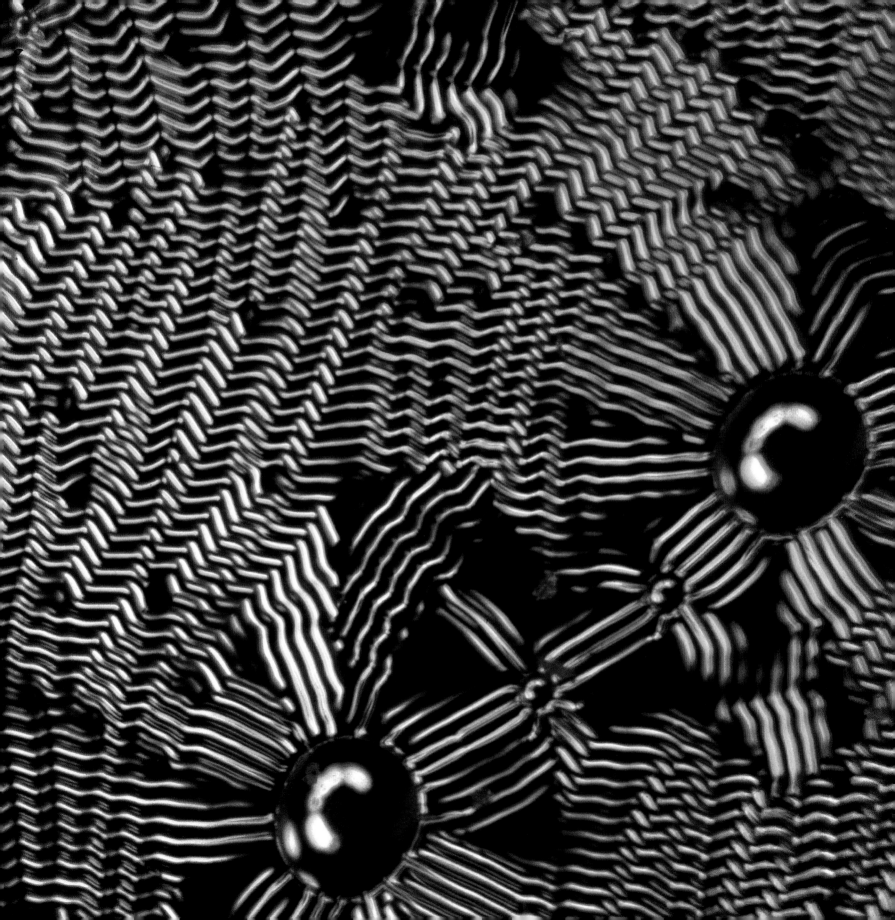

wrinkled gold

PREVIOUS PAGES

A balloon, when full, has a taut, smooth skin; when deflated the skin wrinkles. A ripe grape is like a balloon; dried, it becomes a raisin. A baby is all fat cheeks and smooth curves; his grandfather is a maze of folds and wrinkles.

Here the skin was gold, and the underlying structure a rubbery solid. The gold evaporated from a puddle of very hot, liquid metal formed when an electrical arc — a lightning bolt in a bottle — played over its surface. (This process — "electron-beam evaporation" — sounds dramatic and violent, and it is, but it is a routine operation in making microelectronic devices.) During the condensation of the gold film, the underlying rubber warmed and expanded; when it cooled, it shrank. Wrinkles formed.

The dark-looking balls are gold that exploded off the surface of the puddle of gold as it was heated too rapidly. Because the balls were too thick to expand or shrink, the wrinkles in the more delicate gold film formed around them, like wrinkles at the corners of your grandfather's eyes.

The astonishing regularity of the wrinkles catches the eye. The entire surface is a film of gold — the striped color reflects the angle of lighting and shows the wrinkles.

A log on the fire burns cheerfully; a telephone book smolders and goes out. Both are wood, but they differ in the way they form ash. As the log burns, the ash falls away and exposes fresh wood to the air and the flame. The outside of the telephone book forms a coherent char, which protects the interior by excluding oxygen and heat.

Copper also burns when hot. This pattern is the surface of a copper plate — the bottom of a pan — exposed to the full heat of the cooking flame. The "ash" from this combustion is a protective skin: as the copper combines with oxygen from the air, it forms a thin film of a new compound — copper oxide — on the surface. This film protects the bare underlying copper, at least for a while.

The formation of protective surface films is critical in materials — metals, plastics, ceramics — used at high temperatures. The hot sections of jet turbines and internal combustion engines must be made of temperature-resistant materials; so must cutting tools, bearings, and brake pads. The ability to resist flame by formation of chars is also crucial in fire-resistant building materials.

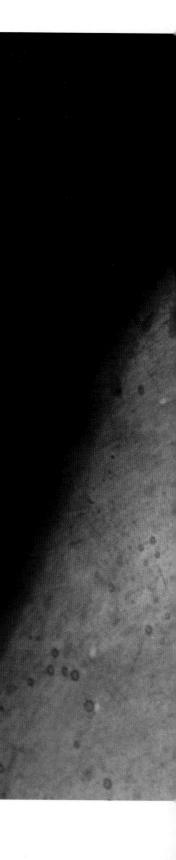

oxide layer
on a copper pan

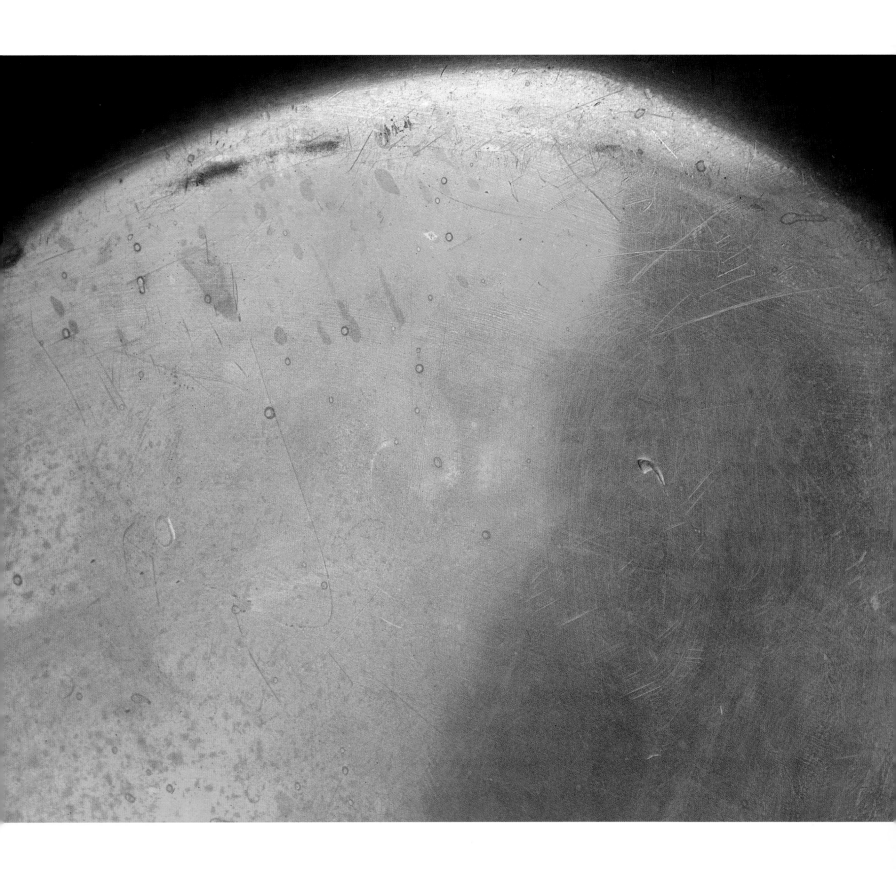

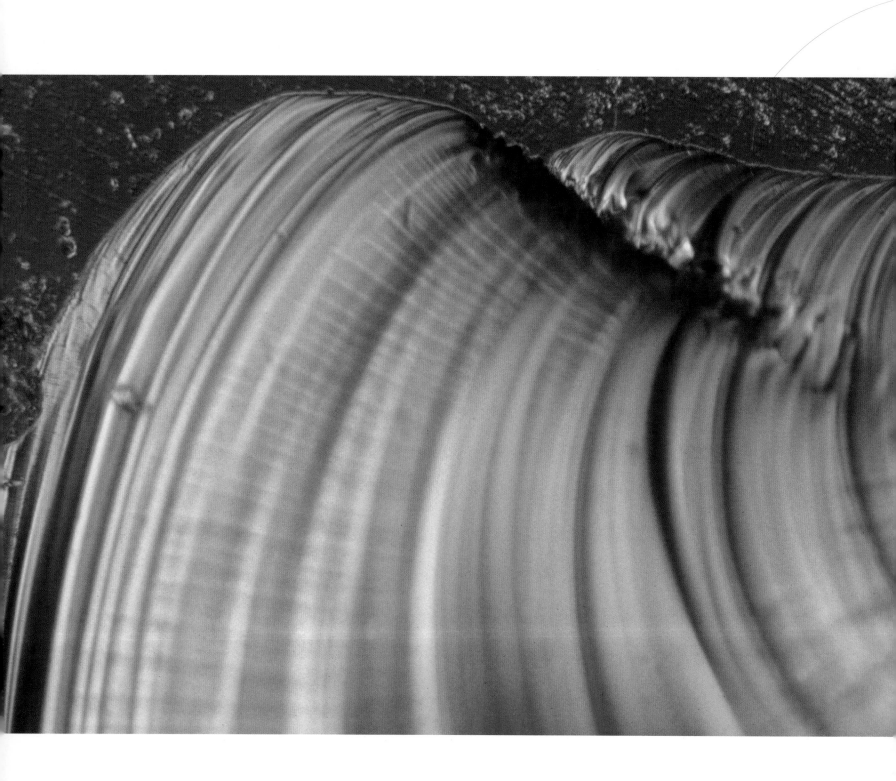

surface of broken glass

It is difficult to tear a handkerchief in half just by pulling on its ends. Notch its edge and it rips easily: the notch concentrates the tension in the threads at the tip of the tear and makes the fabric fail a few threads at a time. Glass fractures the same way: the forces at the tip of the growing crack pull apart the bonds between the atoms of which it is made.

Glass — a three-dimensional network of atoms bonded together — is hard and strong. It is also brittle: glass under tension splinters once a crack starts. When a wine glass hits the floor, it shatters; this apparently instantaneous explosion results from even faster local struggles between the glass and growing cracks. The cracks grow more rapidly than we can perceive, but they have episodic histories: spurts of growth, followed by pauses, recorded on their surfaces as "rings." The tension around a crack concentrates at its tip. As the tension increases, the glass first resists, then fails — a sequence too rapid for us to perceive. The crack shoots forward, ripping bonds apart in sequence — a zipper unzipping at nearly the speed of sound — until the tension is relieved and it pauses. The tension at the tip builds again, until the crack again jumps forward.

Here color helps to map the broken surface. The height of the fracture surface shows undulations that recount these spurts of fracture growth. These ripples are difficult to see, as the glass is transparent. This image uses a specialized microscopic technique to transcribe differences in the slope of the glass surface into colors.

To use solids, it is important to understand how they break. For carbon fiber in a fly rod, strength under tension is crucial; for concrete in the foundations of a bridge and for diamonds in a cutting tool, strength under compression and hardness are required. For automobile bumpers and armor, toughness — the ability to resist a tear or break once it has started — is most relevant.

117

"Sticking" has two parts: the liquid that is sticking and the surface being stuck to. Water is easy to wipe from the floor. Viscous liquids are difficult to remove: honey resists wiping; road tar refuses. The nature of the surface is important, too: nothing sticks to Teflon; everything sticks to an expensive suit.

An adhesive is a viscous liquid (so viscous it may be solid) that sticks well to the surfaces being glued. The most important factor in sticking is the surface. Surfaces that are wetted well by liquids—surfaces whose atoms wish to be covered by other matter—are easy to glue; ones whose surface atoms are indifferent to their exposure are more difficult.

When an adhesive fails—when the glue breaks—it can do so in many ways. The film of liquid adhesive can pull apart, like tearing chewing gum. The interfaces between the adhesive and the glued surfaces can fail. If the adhesive bond is very strong, the objects that are glued together may simply fracture internally.

As this adhesive failed, the very viscous adhesive liquid began to peel back from the surface, splitting into "fingers" that remained stuck to the surfaces, and the adhesive remaining in the fingers moved together laterally, perpendicular to the line of the retreating liquid.

failure in adhesion

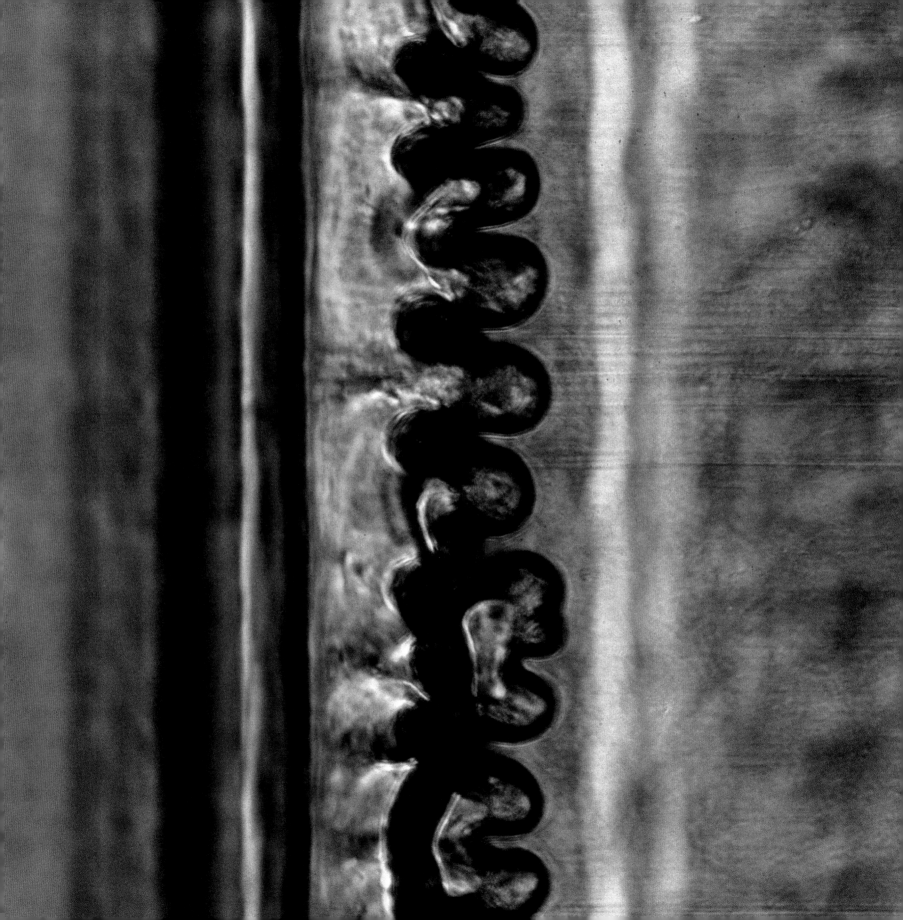

illusion

Our reality is **illusion:**
We don't know for sure
what's out there.

"Reality" is the composite report of sentries. Eyes see;
ears hear; nose smells; tongue tastes; hands touch. Each sends
complicated coded messages to the brain, but consciousness
receives only simplified summaries. Our reality is illusion:
We don't know for sure what's out there.

Sentries sometimes see things that don't exist, and there are
many ways to substitute a mirage for reality. Invisible
things may sometimes also be made visible — a shadow on
an X-ray film describing a tumor; the color of the fire's
embers telling their temperature.

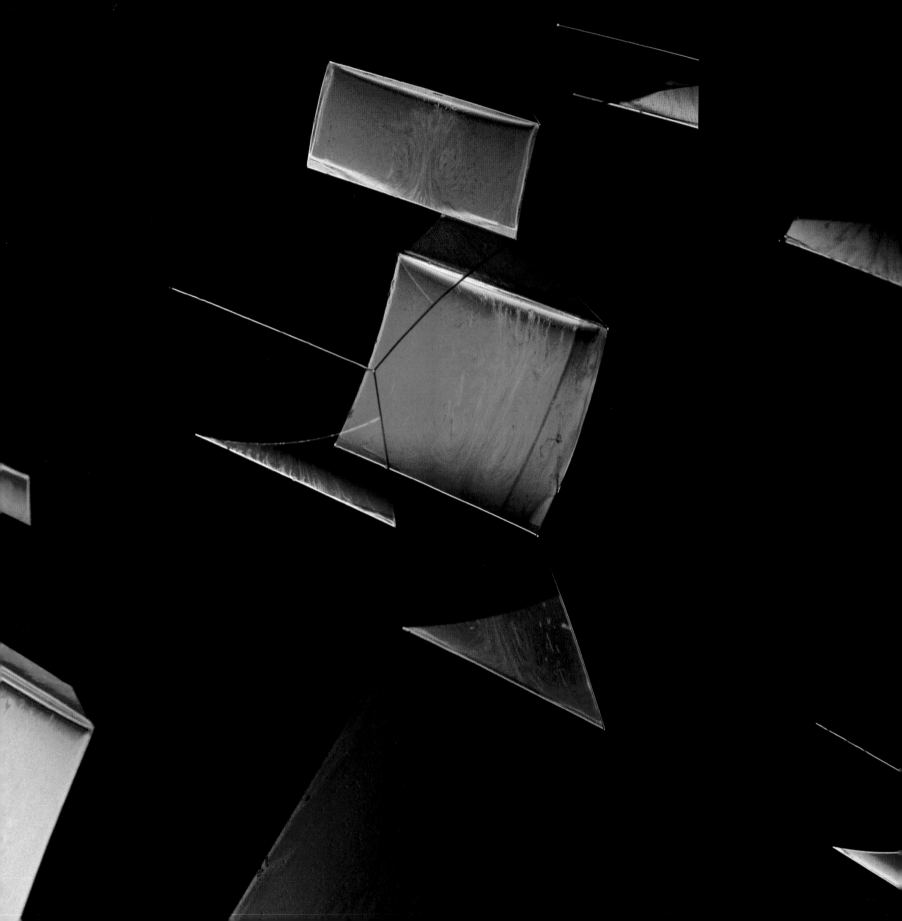

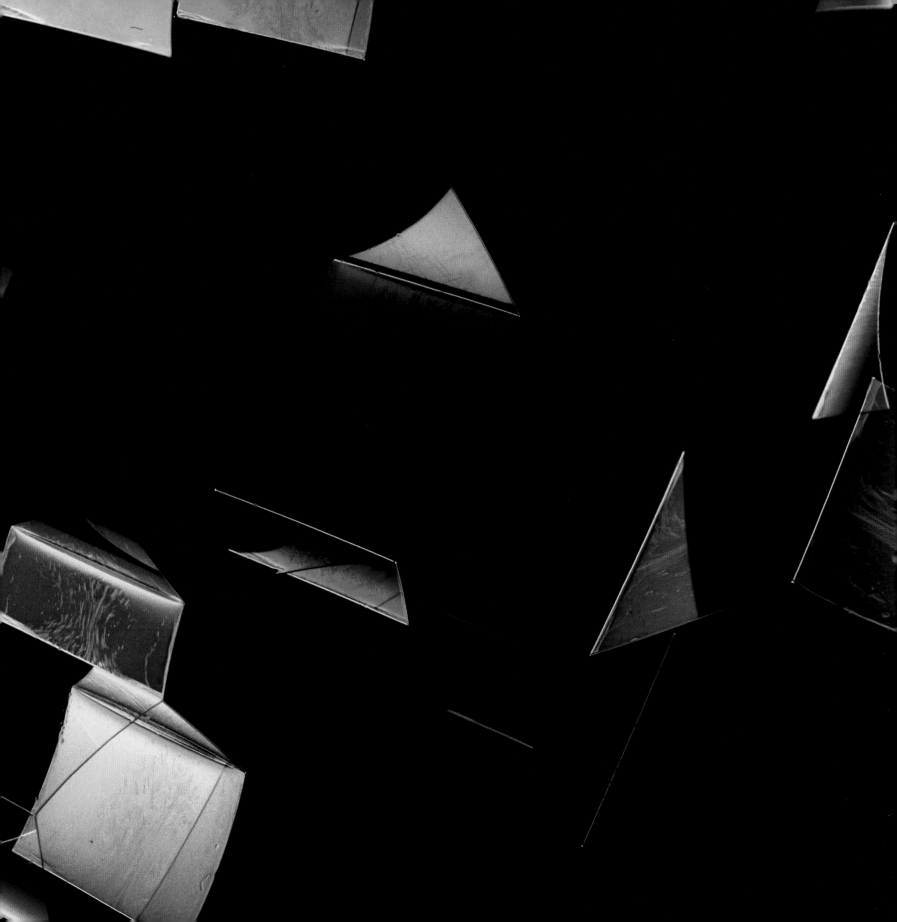

prismatic soap bubbles

PREVIOUS PAGES

Soap bubbles effortlessly answer the question, "For some volume in some space, what is the smallest surface area?" For the bubble blown by a child, the answer is simple: a sphere. For a congregation of bubbles confined between parallel glass plates a few centimeters apart, the answer is a prism.

The sides of these prisms are films only a few thousand molecules thick: the surfaces of the films are soap; their insides are water. Their shape is the one with the smallest area of surface: these films, like all liquids, try to minimize the fraction of their molecules that are exposed at a surface. Most light passes through most of the films without interaction. Occasionally, when the thickness of the film and the size of the light wave are similar, and when the light strikes at the correct angle, a single color reflects. Photons of that color rattle from side to side in the film; some come out on the same side they went in.

These prismatic bubbles were illuminated with white light—a mixture of all colors. Most of the films are invisible; a few reflect. The patterns show the structures of the films with startling precision. The colors define the depths of these microscopic seas, the swirls hint at their currents.

124

A butterfly is an improbable object under the best of circumstances—wings too large to be powered by its tiny body, too defenseless to survive in a hungry world. Its enormous wings are billboards advertising its existence, both to other butterflies and to predators. Each butterfly has a strategy for the use of this space: some paint their wings in camouflage; some color them brightly and advertise for companions; some warn of dire consequences to any eater (like the monarch butterfly, with its load of cardiotoxins extracted from the milkweed on which it feeds). The strategy of some is simply to mimic others' strategies.

When white light strikes the wings of the morpho butterfly from a certain angle, they appear to be a bright, shimmering blue. In fact, they are not blue at all: when viewed from other angles, they are dull browns and grays. The butterfly wastes no metabolic effort in synthesizing blue dye to paint its wings. Instead, its wings are mirrors that reflect only blue light: a collection of transparent microscopic tiles, formed at the thickness required to reflect the color blue selectively. The blue color is pure illusion: when sunlight—white light—strikes the wing, only blue is reflected.

wing of a morpho butterfly

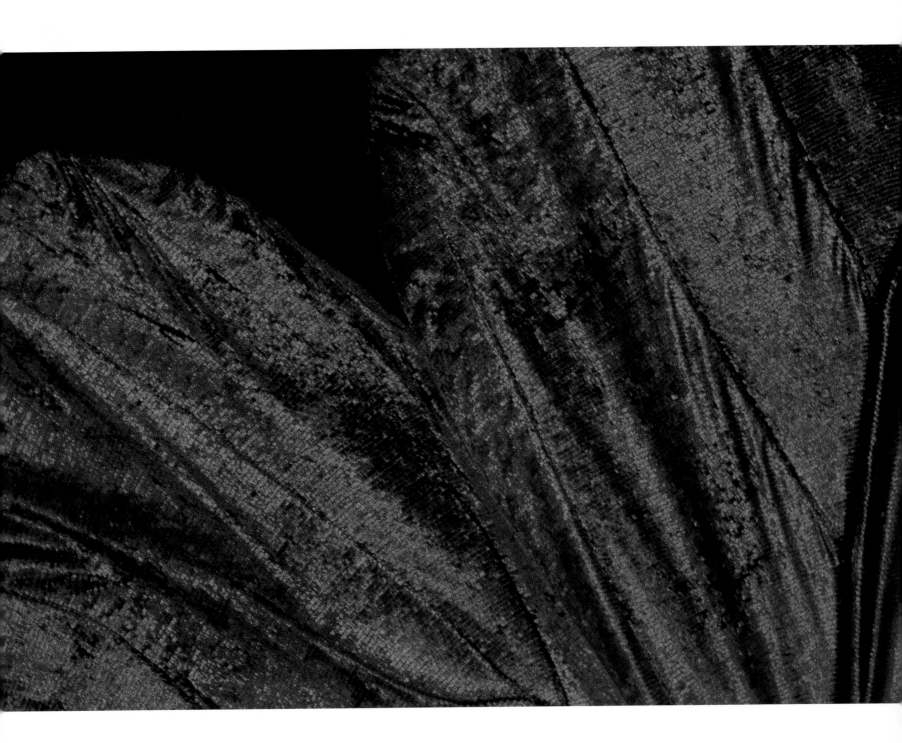

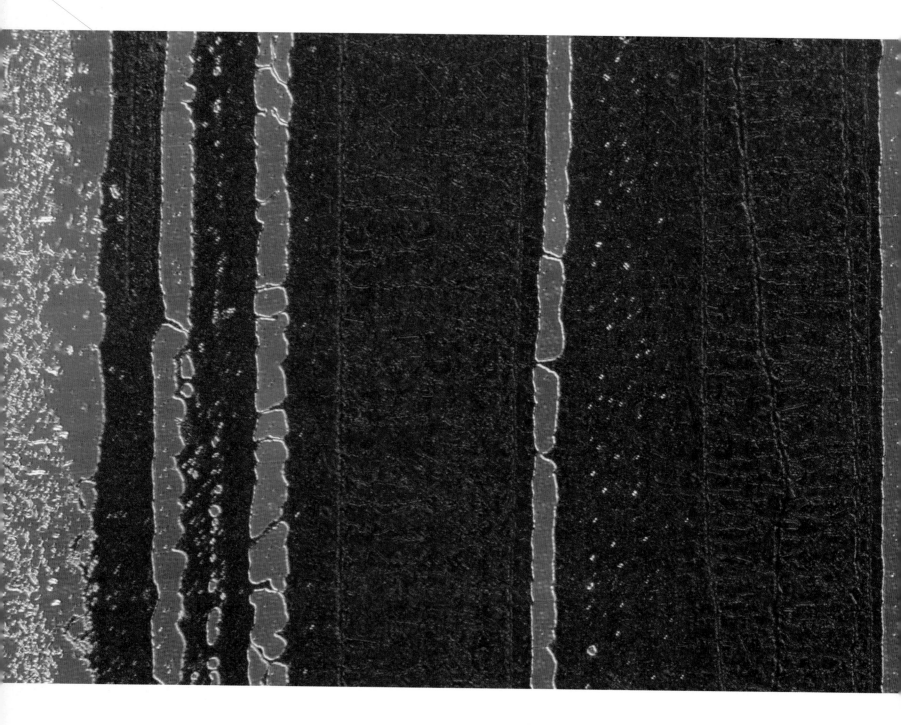

In opal, and in the wings of the morpho butterfly, nature has formed diffraction gratings that assemble themselves. We usually require ruling engines or lasers to draw the countless parallel lines that make up our diffraction gratings.

We can also form such gratings as the opal does, by crystallizing spheres in sheets, or as the morpho butterfly does, by coating a wing with plates of uniform size, but to do so we must first make particles of the correct size. One trick is to use an emulsion—a suspension of tiny droplets of one liquid in another—and to cause the liquid drops, once formed, to solidify into solid spheres. The tiny drops of fat in milk are an emulsion, and the white color of milk comes from the reflection of light from these drops.

By forming plastic microspheres in the liquid droplets of an emulsion, it is possible to make them all one size. Allowing these uniformly sized plastic spheres to settle onto a surface in a thin film produces a regular array, like marbles on a plate. When the spheres are all approximately the wavelength of visible light, the regular thicknesses and spacings of the array make it a diffraction grating.

These images are of a thin film of these crystallized plastic microspheres illuminated with white light. The velvety green color of one, flecked with yellow and red, is the result of diffraction. The second, predominantly brown image, codes the thickness of the ordered layers in colors.

The ability of these structures to appear colored in reflected light when, in fact, they contain no colored dye, suggests new types of paint. Such paints would not lose their color by chipping or fading since there would be no dyed surface layer to fall off or bleach.

colors from
transparent spheres

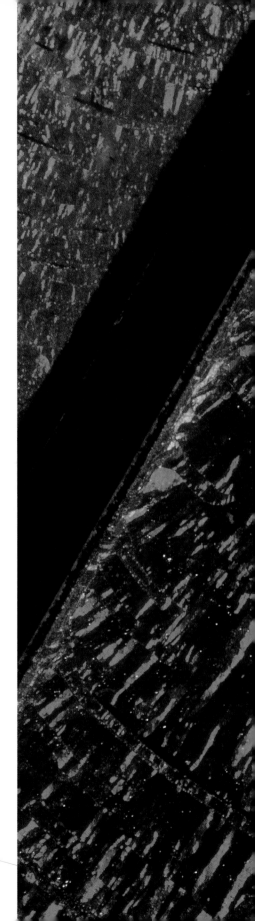

dots on paper

Some patterns can only be seen at a distance: landing lights on the runway at the airport; the spiral arms of a galaxy. The face of your dearest companion is unrecognizable from an inch away.

These dots are a small section of an illustration in a magazine showing the vials of fluorescent particles shown in chapter 1. They are difficult to interpret: the eye is distracted by the extraneous details—the many dots— that do not exist in the object itself. When these dots are reduced in size to the point that they cannot be individually perceived by the eye, the image emerges.

This blending of colors in a pointillist printed image is a blessing. The amount of detail available in a print— the result of fragmenting a recognizable image into several million separate dots—would be overwhelming if our eyes could see it distinctly. The amount of detail in the original object itself is even more extreme: if we could see its individual molecules, we would drown in useless information.

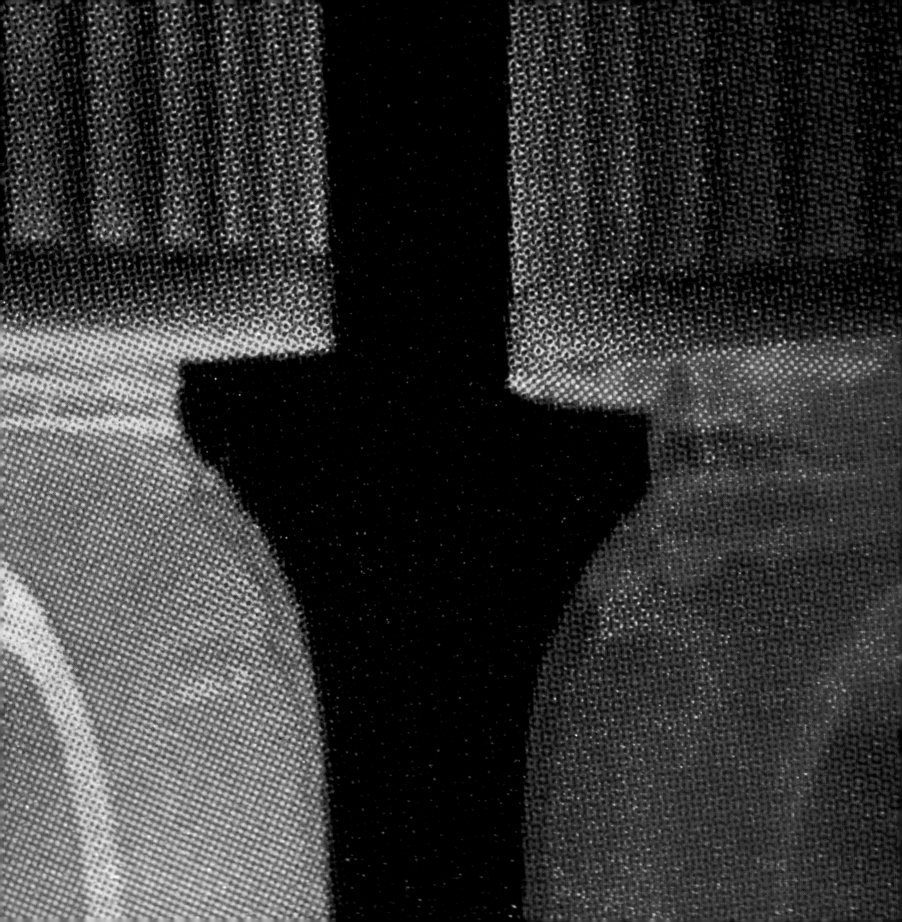

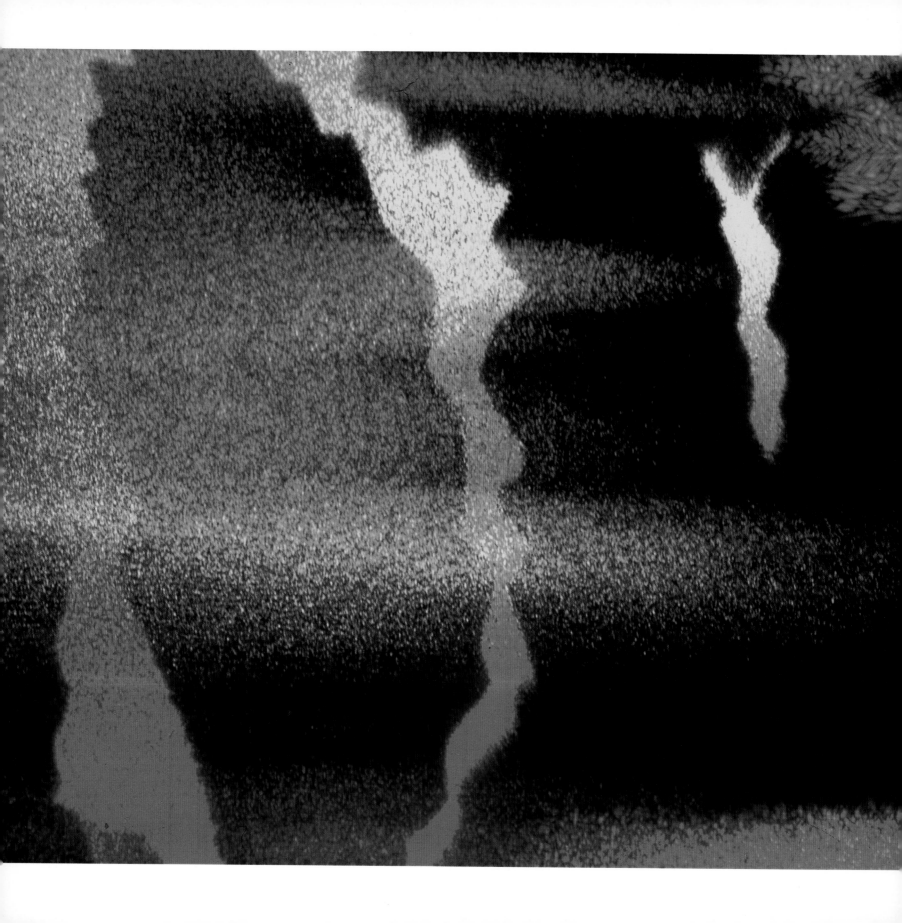

h o l o g r a m

"Seeing" is a kind of shadow play of images projected onto the screen of the retina. Light—high frequency electromagnetic ripples—strikes a coffee mug. The surface of the mug responds, reflecting some of the light waves. The lens of the eye focuses these reflected ripples onto the retina, where the excitation of cells in the retina is first interpreted locally, then actively discussed in a dialogue between the retina and the brain, and finally transmitted to the brain as a series of pulses in the optic nerve. Using this information, and taking clues from differences between the reports from the two eyes, the brain infers the mug's three-dimensional shape. Any process that builds the pattern of light that our eyes and brain interpret as "mug" will be a mug to us. We have no way of knowing if the patterns of light that fall on the retina are real or counterfeit without independent checks like smelling, feeling, and hearing.

With your eyes closed, a good stereo system is difficult to distinguish from a live orchestra: the sounds—the ripples of pressure in the air—produced by the two are indistinguishable. You can walk around in the sound from the stereo, and the pattern of vibrations is the same as in the symphony hall. Holography performs a similar trick for sight, recreating three dimensions from two. It uses interference among waves of light reflecting from the rippled surface of the hologram to build a three-dimensional pattern of light in space very similar to that coming from a mug. You can move your head and look around in this pattern, seeing the same variations in it that you would see with a real mug.

The major present uses of holograms are to provide security labels for credit cards and paper currency. In the future, holography may present three-dimensional information in books and on computer screens, and be a part of virtual reality.

131

Wave an electrical charge at an electron, and the electron waves back. Like charges repel; unlike attract. If the frequency of waving is sufficiently high (a million billion times per second), the interaction is "light." The electrons in metals are especially responsive, since they are relatively loosely held by their nuclei. Shine light on a very flat surface of a metal (a mirror), and the electrons in the surface oscillate in response; their collective oscillation regenerates the light. We call this regenerated light "reflection." Reflections can appear indistinguishable from originals when the mirror is sufficiently flat.

But what about the properties of this reflection? We look in a mirror, and—at first glance—everything seems reversed right to left, but, astonishingly, not reversed top to bottom. We hold up our right hand—the one with the ring on the thumb—and the image holds up its left; the symbol ▶ appears to be ◀ in the mirror. Right to left.

What we see and what we interpret are, however, not necessarily the same. We say that our right hand is the image's left hand because we have

imagined walking through the plane of the mirror and standing in the place where we imagine the image to be standing, facing us. To stand there, in that orientation, would require us to turn around. *If* we were to walk through the mirror and *if* we were to turn to stand as our mirror image stands, the hand that moved would be our left . . . but wait. The hand of the image that moved was the hand that had the ring on it. Has the mirror both changed right and left and also moved the ring from one hand to the other? What *is* right, and what *is* left, in the mirror?

What we see in the mirror is a complicated gavotte danced by our eyes and our brain, and usually not to the same tune. The eyes simply do the optics; the brain superimposes assumptions on the interpretation without telling anyone. The result is

confusion. The mirror does not uniquely interchange right and left, leaving top and bottom unchanged. It is probably most accurate to say that it interchanges front and back. Really *not* easy to understand.

Reflection in the mirror suggests how dependent our sense of reality is on assumptions that we make without knowing that we are making them.

reflections   ○

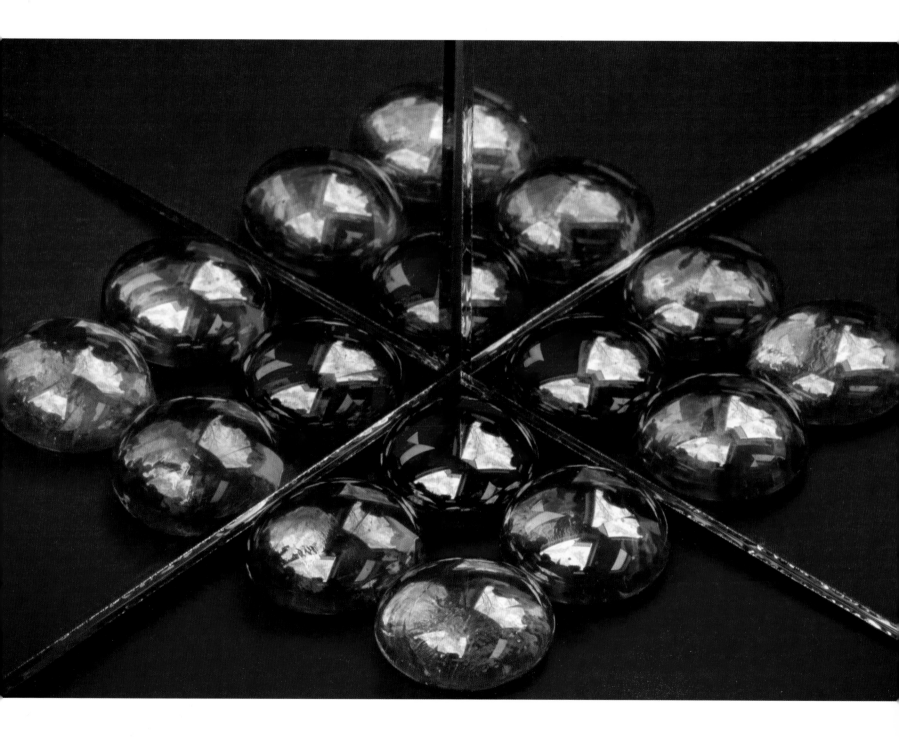

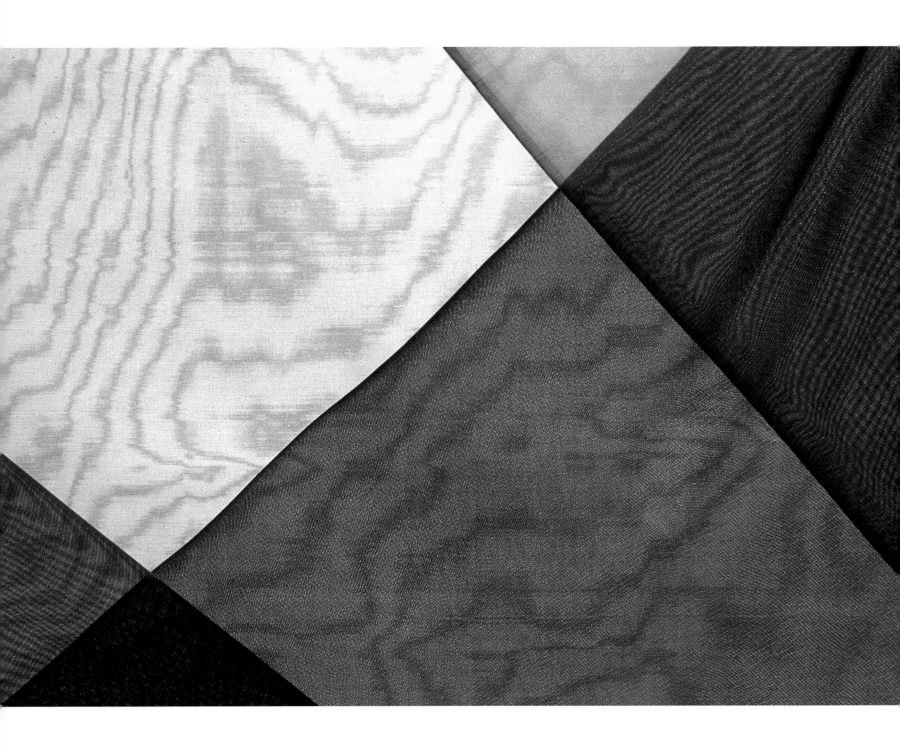

In late-night reruns of old Westerns, the wagon wheels often turn backward. The shutter of the camera blinks at a fixed rate, each blink freezing the position of the spokes. Depending on the rate of the shutter and the speed at which the wheel turns, the spokes may appear to turn in reverse.

Differences between two periodic events or structures are common. Two penny whistles playing the same note will warble: the frequencies of the note will be slightly different for the two, and the waves of sound from them will periodically come into step and go out of step. The patterns here—moiré patterns—come from overlapping lay-

moiré pattern ○

ers of fabric. As the patterns of threads come into and go out of coincidence, subtle patterns of new lines appear. These lines have a wavelength much longer than that between the threads.

Moiré patterns convert small differences between two repeating patterns into another pattern with a much longer period of repetition. This third pattern can be used to make very precise measurements of distance and movement, by allowing two periodic structures to move relative to one another and detecting the difference between the two. Such systems measure dimensions of ultraprecision parts and optical components like lenses, and magnitudes of small movements of structures, when they are deformed mechanically.

light reflecting from a polymer-containing gel

Imagine the end of an evening: an expensive restaurant, an elegant but almost empty wineglass, a few strands of spilled tagliatelle, a half-finished baklava. Also, boredom. The evening has gone on too long. A way of passing the time: dip your finger in the wine, and gently rub the rim of the wineglass. You—or anyone—can generate a humming note that sounds a little like a fat housefly frustrated by a windowpane. If the dinner is *really* ending badly, several people may try it simultaneously, with the result sounding like an untuned glass harmonica.

The vibration on the fingertip tickles; it's a mechanical resonance. The gentle friction of the fingertip along the rim makes the glass vibrate. Any particular glass on the table will produce only one note. It may get louder if rubbed harder, but it does not change pitch: the vibrating glass is a kind of bell.

The colors of this circular pattern are also resonances, but of light, not sound: electromagnetic waves, not pressure waves in air. In this experiment, the "wineglass" does not look anything like a wineglass—in fact, if anything, it resembles the baklava. What you see is a colorless gel made of a particular kind of polymer dissolved in a solvent. A molecule of polymer is usually like a piece of tagliatelle: a limp, uniform length of molecular string. Here, the two ends were slightly different—one end stickier, the other stiffer (an innovation perhaps still unexplored in Italian cuisine). This particular polymer forms itself in a layered structure, with the stiff ends assembling into sheets to form the "pastry" of the baklava, and the sticky ends the "honey." The layers are spaced much too closely together (a few hundred nanometers separation) to see by eye, but they have the right dimensions to interact strongly with visible light.

But why the circular patterns? The structure was formed when a small quantity of the solution of polymer was placed under a thin, circular glass disk, a few centimeters across. The spacing between the layers of molecular baklava—that is, the layers of self-assembled polymer—increased slightly from the center to rim of the disk, for reasons that depend on molecular details of the experiment. White light passing through the subtly changing structure made its different parts "hum" at the frequencies of different colors. Here, surprisingly, rings of different color reflect back toward the source of light. (Although the colors are very different to our eye, their frequencies are similar—from "red" to "blue" is a change of less than a factor of two.)

You can think of this system as a kind of three-dimensional diffraction grating that obligingly forms itself from a solution of loosely linked, transparent, colorless molecular spaghetti. These gels form themselves into structures that selectively reflect specific colors—a lovely characteristic, and one that they share with the peacock's magnificent tail. They might make colorful paints or inks.

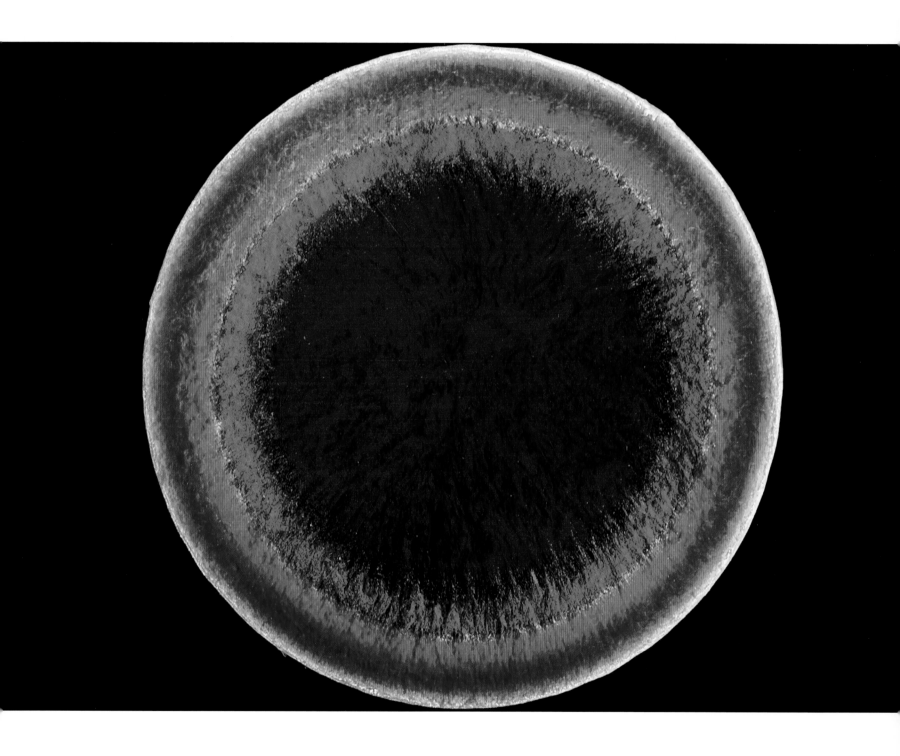

notes and readings

PHOTOGRAPHIC NOTES BY FELICE FRANKEL
TECHNICAL NOTES BY GEORGE WHITESIDES

George and I decided to use the opportunity of a new edition to replace some of the images and text from the original printing. The images and prose of the first edition had aged astonishingly well, but we still wanted to introduce new ideas.

At the time of that first printing, I was asked to submit original film for color separations. Some of the images were first scanned and digitally "cleaned" (see below), and then recorded again on 4×5 film before being separated (a technique used for illustrated publications at that time). An interesting problem emerged in converting the original images into digital form for this new submission. I had hoped to retrieve the original scans. I contacted the lab that had scanned the 35-mm slides, and to my amazement the lab had kept the digital files. However, to my intense frustration, there was no way to read the files! The technology had changed so drastically that a usable digital reader for the files was nowhere to be found.

Those of us who archive images and other files must ask how we are to read these files years in the future? Something to think about.

And so, for this reprinting, I scanned all the images again, on my own scanner this time, and sometimes "cleaned" the files so that the reader would not be distracted from the important ideas in the images. The issue of cleaning science images is an ongoing concern within the science community; I discuss this issue in my book *Envisioning Science: The Design and Craft of the Science Image,* in which I address the differences between images for journal submissions and books for the public. I believe I am not crossing any ethical lines by digitally manipulating an image; I always tell the reader when I have done so.

I used a Nikon F3 unless otherwise specified. I usually shot at f 22 for the nonmicroscopic images and bracketed half stops up and down a full stop. Shutter speeds ranged from 1/8 second to 2 minutes. Film choices were Velvia 50 and Ektachrome 64 (tungsten). Many of the italicized photographic notes that follow are accompanied by shorthand designations to indicate details of their production.

The macroscopic images (Ma) were taken with either a 105 mm or a 55mm macro lens, always on a tripod. The stereomicroscopic (SM) and microscopic (Mi) images were taken with camera adapters attached to each scope. Scanning electron microscopy (SEM) was usually performed with the help of a researcher. Digital cleaning is indicated with a (Cl) and digital coloring with (DC). I used epi-illumination for most of the microscopic images and used Nomarski Differential Contrast (N) to emphasize surface structure. I give credit whenever possible to those who fabricated samples or helped with photography.

Technical notes by George Whitesides and references that follow each photographic note provide more information about the details of the experiments and an introduction to the scientific literature.

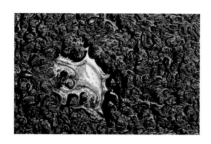

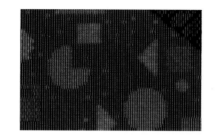

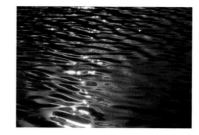

Oil Slick (page 15)

After hosing down my driveway on a gray Sunday afternoon I searched for one of the more interesting puddles and dropped a bit of oil onto it. I waited half an hour until the diffraction colors became interesting. With the camera and 105 mm lens on my tripod I shot perpendicularly to the slick.

The bands of color are due to constructive interference between light reflecting from the air-oil and oil-water interfaces. They occur when the distance the light travels inside the oil film is a multiple of the wavelength of a color of light in that film. As the film becomes thicker, the colors therefore re-emerge.

Minneart, M. G., *Light and Color in the Outdoors*, L. Seymor, trans., Springer Verlag, New York (1993).

Ramme, G., "Reflected laser light from a soap bubble—a demonstration experiment," *Physics Education*, 27 (1992), 282–86.

Smith, F. G., and J. H. Thomson, *Optics*, 2nd ed., John Wiley & Sons, New York (1988).

Computer Monitor Screen (pages 16–17)

Although somewhat dated, this Pacman™-like design on the screen of my Mac PowerPC 7600™ from ten years ago still tells a relevant visual story of how a computer screen appears, up close. I set the tripod in front of my computer screen and used a 2x extender on my 105 mm macro and shot with daylight film. Ma

White, R., *How Computers Work*, 8th ed., Pearson Education, Que Publishing, Indiana (2005).

Reflections of Sunlight on Water (page 19)

I took this picture while photographing the garden at Stourhead in England. Using my 105 mm lens, I found this particular exposure was best for emphasizing the sun's reflections on the lake. Using color slide film for what is basically a black and white image gives the quality I was looking for.

When light encounters an interface between two media with different indices of refraction (e.g., between air and water), the amounts of light that pass across that interface and reflect from it are a complicated function of the angle of incidence, the indices of refraction, and the polarization of the light. In general, the larger the difference in index of refraction, the more light reflects.

France, M. M., "Reflections on rippling water," *American Mathematical Monthly*, 100 (1993), 8, 743–48.

Lynch, D. K., and W. Livingston, *Color and Light in Nature*, 2nd ed., Cambridge University Press, Cambridge (2001).

139

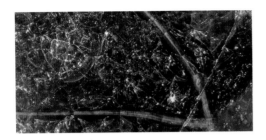

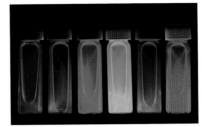

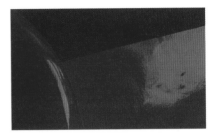

140

Veins of Opal (pages 20–21)

William Metropolis, Assistant Curator of the Mineralogical Museum at Harvard University, was kind to let me borrow a few large samples of opal. I used daylight for this particular specimen and captured the diffracting light with my 55 mm micro lens. Ma

Opals are crystals of spherical colloid particles (usually of silica) embedded in a matrix with a different index of refraction. The colors that reflect from opal are determined both by the size of the colloidal spheres and the orientation of the lattice.

Murray, C. A., and D. G. Grier, "Colloidal crystals," *American Scientist*, 83 (1995) 238–45.

Ohara, P. C., et al., "Crystallization of opals from polydisperse nanoparticles," *Physical Review Letters*, 75, (1995) 3466–69.

Suspensions of Small Fluorescent Particles (pages 22–23)

Laying the vials horizontally to add air bubbles to the composition, I photographed the fluorescing samples under ultraviolet light with a 380 nm filter to absorb extraneous UV. At the time, I found Velvia to be the best choice for film. The material was from the work of MIT professors Moungi Bawendi and Klavs Jensen, and Christopher Murray, Cherie Kagan, Bashir Dabbousi, and Javier Rodrigez-Viegothen, then graduate students at MIT.

Cadmium selenide nanocrystals were prepared by precipitation and a monolayer of organic material was attached to the outside of the crystals, with three useful effects: the organic layer acted as lubricant to prevent the particles from coagulating irreversibly in suspension; it protected their surfaces from reaction with oxygen and water; and it controlled the separation between particles when they crystallized. The ability to change both the electronic structure of individual particles and the electronic coupling between them makes this system attractive for studying the electronic structure of quantum dots. These particles are increasingly interesting in materials science because they bleach only very slowly on exposure to light, and because the same process can make particles of different sizes, and thus with different colors of fluorescence.

Murray, C. B., Kagan, C. R., and Bawendi, M. G., "Self-organization of CdSe nanocrystallites into three-dimensional quantum dot superlattices," *Science*, 270 (1995), 1335–37.

The Color Red (pages 24–25)

Here, I composed an edge of a ceramic platter (by artist David Hayes) and two red papers under my office skylight. I used my 55 mm macro lens. An ink mark on one of the papers was digitally removed. Cl

These surfaces are "red" in all directions because their surface is microscopically rough: the light that reflects from the surface is reflected in all directions (that is, it is "scattered").

Chamberlin, G. J., *Colour, Its Measurement, Computation, and Application*, Heyden Publishing, London (1980).

"It's a Colorful Life": http://www.sci-ed-ga.org.

Schopenhauer, A., *On Vision and Colors*, Berg Publishers, Oxford (1994), originally published in 1816.

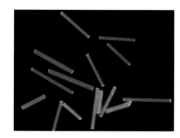

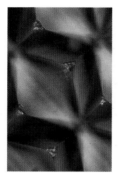

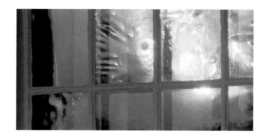

Tubular Gels (page 27)

The late Professor Toyoichi Tanaka of the Department of Physics at MIT was a friend and colleague. He asked graduate student Changan Wang to create these gel shapes at my request. They were divided into three batches, which were then placed in solutions of three different fluorescing dyes. I composed the rods on a black tray and lit them with UV light. The gels that absorbed one of the dyes, rhodamine B, did not read well on tungsten film and so I digitally tinted them orange to approximate what my eyes saw. Ma, DC

The gels are cross-linked polyacrylamide and were made by polymerization of acrylamide monomer in a glass tube, followed by extrusion of the rod of gel from the tube. The dyes are fluorescein (green), ethyl courmarin (blue), and rhodamine B (red). This image represents a solution to a conundrum: what colors do you choose when two recording media (the human eye, and photographic film) have different sensitivities to color, and hence "see" different colors? The decision to tint the image was an effort to replicate, in print, the response of the eye.

Matsuo, E. S., and T. Tanaka, "Patterns in shrinking gels," *Nature*, 358 (1992), 482–85.

Retroreflectors (page 28)

William Bernett, formerly at 3M, supplied me with a sample of the material. I photographed it using a combination of reflected and transmitted light. Mi

This array of corner cubes is a 3M product and is part of a broad technology developed by this and other companies based on microreplication. This material is used for optical retroreflectors. Retroreflectors are structures that reflect light back toward its origin. They are widely used in (for example) road signs and bicycle reflectors, where the objective is to reflect light from the headlights of an automobile back to the driver of the automobile. They are also used to enhance the brightness of computer displays by directing more of the light toward the user, to distribute light in illumination systems, and to produce abrasives in which the shapes and orientations of the surfaces used in cutting are precisely controlled.

Gagliardi, J., and T. Buley, "Study of STI polishing defects using 3M fixed abrasive technology," CMP-MIC Proc., Mar. 9, 2001, pp. 535–38.

Parker, D. H., et al. "Attenuated retroreflectors for electronic distance measurement," *Optical Engineering*, 45 (2006), 7.

Sun, Reflected (pages 30–31)

In my former life as a landscape photographer, I was given the memorable assignment to photograph sculptor Augustus Saint-Gaudens' garden, in Cornish, New Hampshire. As the sun was setting and as my shutter speed settings began increasing, I turned my attention to the glorious reflection on the parlor window leading out onto the porch.

Fleming, R. W., R. O. Dror, and E. H. Adelson, "Real-world illumination and the perception of surface reflectance properties," *Journal of Vision*, 3 (2003), 347–368.

"Optical Illusions and Visual Phenomena": http://www.michaelbach.de/ot.

Sekal, A., *Incredible Visual Illusions (You Won't Believe Your Eyes!)*, Firefly Books, Toronto (2004).

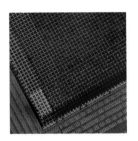

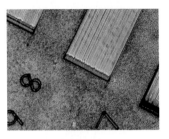

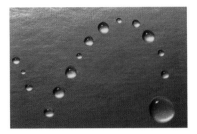

Focal Plane Array

(pages 32–33)

David Hitrys at QImaging Corporation sent me this ICX282AK digital camera sensor, made by Sony. I attached my Nikon digital camera to the microscope and later "nudged" the histogram just a bit. Some of the Bayer filter mosaic is apparent. The latter refers to a particular arrangement of color filters used in most single-chip digital image sensors to create a color image. The filter pattern is 50 percent green, 25 percent red, and 25 percent blue. Mi

This structure is a complementary metal oxide semiconductor (CMOS) chip. This type of chip uses a mixture of transistors with negative and positive polarity, and in doing so conserves energy and simplifies design. CMOS is the technology that now dominates the fabrication of complex devices fabricated in silicon. A focal plane array is a planar array of pixels, each of which records the intensity of light that falls on it and stores that intensity; these intensities are read by the digital camera and provide the basis for the image that it produces.

Editorial, "Seeing science in color," *Nature— Structural & Molecular Biology*, 14 (2007), 173.

Minkel, J. R., "Camera reconstructs image from single pixel," *Scientific American*, October 6, 2006.

Silicon, Etched by Light

(pages 34–35)

Theodore Bloomstein's work for his doctoral thesis in the Department of Electrical Engineering and Computer Science at MIT involved laser etching silicon surfaces. I used his reference numbers as a compositional element. Mi

With this technology, three-dimensional structures can be rapidly machined in silicon to micrometer tolerances. This work benefits the emerging field of microelectromechanical systems (MEMS) by extending milling techniques to new size regimes. MEMS merges electronic and mechanical components to produce sensors and actuators such as pressure transducers, valves, and pumps. Current techniques for fabricating these mechanical components rely on lithographic techniques that limit shapes to extrusions of two-dimensional patterns. Laser processing offers a means to produce mechanical components that are inherently more three-dimensional.

Bloomstein, T. M., and D. J. Erlich, "Laser deposition and etching of three-dimensional microstructures," TRANSDUCERS '91: 1991 International Conference on Solid-State Sensors and Actuators, Digest of Technical Papers, 507–11.

———, "Stereo laser micromachining of silicon," *Applied Physics Letters*, 61 (1992), 708.

Madou, M. J., *Fundamentals of Microfabrication: The Science of Miniaturization*, 2nd ed., CRC Press, Boca Raton, 2002.

Drops of Water (page 38)

I individually placed the drops with the tip of a steak knife on a gold-coated fiftieth anniversary card. Ma

These drops are water drops on a hydrophobic plastic surface. The shadowing of the drops shows internal lensing. The surface of the drop focuses the light on the surface opposite to that on which the light falls. This lensing can have practical consequences: drops of water on the surface of a car, in the bright sun, can focus the sunlight on the paint and cause local damage.

Adamson, A. W., and A. P. Gast, *Physical Chemistry of Surfaces*, 6th ed., John Wiley & Sons, New York (1997).

Kim, E., and G. M. Whitesides, "The use of minimal free energy and self-assembly to form shapes," *Chemistry of Materials*, 7 (1995), 252–54.

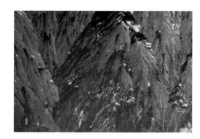

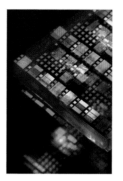

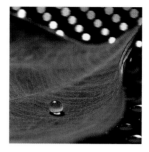

Ice Crystals (pages 40–41)

I pointed my camera through an ice-covered window into the winter sunset using daylight film. Ma

These quasi-two-dimensional crystals of ice are growing in kinetically controlled patterns and are strongly influenced by mass transport (and possibly heat transport).

Davey, R., and D. Stanley, "All about ice," *New Scientist,* 140 (1993), 33–37.

Harrison, A., *Fractals in Chemistry,* Oxford Science Publishing, Oxford (1995).

Libbrecht, K., *Ken Libbrecht's Field Guide to Snowflakes,* Voyageur Press, St. Paul (2006).

Walker, J., "Icicles Ensheathe a Number of Puzzles: Just How Does the Water Freeze?" *Scientific American,* 258 (1988), 114–17.

Stamp for Microprinting
(page 42)

Tungsten lamps produced the diffraction colors for former postdoctoral fellow James Wilbur's and Rebecca Jackman's sample from the Whitesides lab. A similar image appeared on the cover of Science, *August 4, 1995. Ma*

An elastomeric stamp, made of transparent polydimethylsiloxane (PDMS), is formed by casting the polymer against a master structure that has a pattern of features in bas-relief on its surface. The stamp is "inked" with a compound that forms a self-assembled monolayer on the surface being patterned: a typical combination is a PDMS stamp, an alkanethiol "ink," and a flat gold surface. The inked stamp comes into contact with the gold surface; an image complementary to the pattern on the surface of the stamp forms. The resolution of this pattern is remarkable: the roughness of the edges of the pattern is less than 50 nm, even when the process is carried out in the open laboratory. The colors from the front side are those of diffraction of light propagating in air; those from the back side are from diffraction of light propagating in PDMS.

Love, J. C., et al., "Self-assembled monolayers of thiolates on metals as a form of nanotechnology," *Chemical Reviews,* 105 (2005), 1103–69.

Whitesides, G. M., J. K. Kriebel, and J. C. Love, "Molecular engineering of surfaces using self-assembled monolayers," *Science Progress,* 88 (2005), 17–48.

Nonwetting Surfaces
(page 45)

Rob Nicholson from the Botanic Garden at Smith College gave me a number of leaves to play with.

http://lotus-shower.isunet.edu/the_lotus_effect.htm.

Otten, A., and S. Herminghaus, "How plants keep dry: A physicist's point of view," *Langmuir,* 20 (2004), 2405–08.

Zhai, L., et al., "Stable superhydrophobic coatings from polyelectrolyte multilayers," *Nano Letters,* 4 (2004), 1349–53.

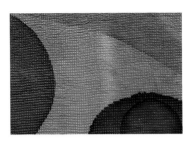

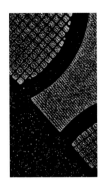

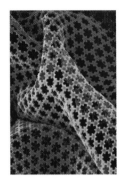

144

Inks Bleeding on Fabric
(page 46)

This is a detail of one of my favorite handmade scarves. The artist had placed the inks on the fabric anticipating how far they would bleed. I shot it under diffused daylight with the 55 mm lens.

Capillary wicking is almost universal on contact of a liquid with a porous solid, so long as the contact angle of the liquid on the solid is less than 90°. The liquid-vapor interfacial free energy may or may not be important in capillary wicking: if the capillaries are completely closed, wicking may result in very little change in liquid-vapor surface area. In wicking of liquid into this porous fabric, there is a significant increase in the liquid-vapor area, and the surface tension of the liquid is important.

Adamson, A. W., and A. P. Gast, *Physical Chemistry of Surfaces*, 6th ed., John Wiley & Sons, New York (1997).

Isrealachvili, J. N., *Intermolecular and Surface Forces, with Applications to Colloidal and Biological Systems*, 2nd ed., Academic Press, New York (1992).

Patterned Crystals (page 47)

Joanna Aizenberg, now Professor of Materials Science at Harvard University, gave me a number of her fascinating samples. I made this image with darkfield microscopy. Mi

Aizenberg used microcontact printing to form a pattern of a carboxylic acid-terminated self-assembled monolayer on a thin gold film (supported on a silicon wafer). The spaces between the regions printed with the carboxylate-terminated self-assembled monolayers (SAMs) were filled with a hydroxyl- or ethylene glycol-terminated SAMs. On exposure to a solution of calcium carbonate, under conditions in which calcite nucleation occurred, crystals formed on the carboxylate-terminated regions. This work is an example of "materials by design": that is, the design of *molecules* that assemble in a way that yields a *material* with specific, macroscopic properties.

Aizenberg, J., A. J. Black, and G. M. Whitesides, "Oriented growth of calcite controlled by self-assembled monolayers of functionalized alkanethiols supported on gold and silver," *Journal of the American Chemical Society*, 121 (1999), 4500-09.

Weibel, D. B., W. R. DiLuzio, and G. M. Whitesides, "Microfabrication meets microbiology," *Nature Reviews Microbiology*, 5 (2007), 209-218.

Molded Plastic Microfabric
(page 48)

Younan Xia, now a Professor of Chemistry at Washington University in St. Louis, took the original gray-scale scanning electron microscopy (SEM). I digitally colored the image using Adobe Photoshop. The colored image appeared on the cover of Nature, *August 17, 1995. SEM, DC*

A transparent polydimethylsiloxane (PDMS) stamp was prepared by casting a liquid prepolymer across a relief structure of photoresist; the pattern was introduced by conventional photolithography. The PDMS stamp, on contact with a flat piece of glass, formed a network of channels with minimum channel diameter of approximately 1 µm. A liquid epoxy prepolymer was brought into contact with this network and filled it by capillarity. Thermal or photochemical crosslinking of the prepolymer (using light transmitted through the PDMS) gave a solid network. Peeling away the PDMS stamp left the network of polymer on the glass, which was then dissolved to release the plastic fabric.

Gates, B. D., et al., "Unconventional nanofabrication," *Annual Review of Materials Research*, 34 (2004), 339-72.

———, "New approaches to nanofabrication: Molding, printing, and other techniques," *Chemical Reviews*, 105 (2005), 1171-96.

Kim, E., Y. Xia, and G. M. Whitesides, "Polymer microstructures formed by moulding in capillaries," *Nature*, 376 (1995), 581-84.

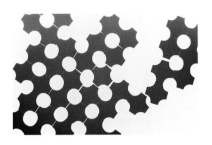

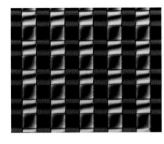

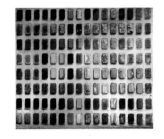

Self-Assembled Structure

(page 51)

It was remarkable how quickly the structures assembled as soon as I placed them at the interface of perfluorodecalin and water. Graduate student Ned Bowden in the Whitesides lab generated the individual shapes. I photographed them in a round glass laboratory dish over white paper under daylight with a 105 mm lens. Ma, Cl

Small, cross-shaped pieces of polydimethylsiloxane (PDMS) were prepared in several steps. A long rod of PDMS with this cross section was cast from a mold. The flat sides of the rod were covered with transparent tape, and the exposed faces of PDMS were oxidized in a plasma generator until they became hydrophilic. The tape was removed from the PDMS, and the individual pieces of PDMS were cut from the rod and floated between perfluorodecalin (chosen for high hydrophobicity and density) and water. The hydrophilic polymers were wetted by water; the perfluorodecalin wet the hydrophobic surfaces, creating menisci. The elimination of these menisci caused the assembly.

Bowden, N., et al., "Self-assembly of mesoscale objects into ordered two-dimensional arrays," *Science*, 276 (1997), 233–35.

Whitesides, G. M., and Boucheva, M., "Beyond Molecules: Self-assembly of mesoscopic and macroscopic components," *Proc. Natl. Acad. Sci. USA*, 99 (2002), 4769–74.

Whitesides, G. M., and B. Grzybowski, "Self-assembly at all scales," *Science*, 295 (2002), 2418–21.

Coated Silver Wires

(pages 52–53)

Paul Nealey, a postdoctoral fellow in the Whitesides lab, produced the sample. The criss-crossing of the polymer created interference patterns exaggerated with Nomarski. Mi, N, Cl

A substrate was prepared by evaporation of a 50-nm thick film of silver onto a silicon wafer. A pattern of 50-μm-wide lines was printed onto the surface using microcontact printing. The surface not covered with this initial self-assembled monolayer was covered by a resist by dipping the initially patterned surface into an appropriate solution. Dipping the patterned surface into liquid prepolymer (an optical adhesive) resulted in covering the hydrophilic parts of the pattern with liquid prepolymer and leaving the hydrophobic parts bare. The polymer was cured by exposure to UV light, and the resist and the exposed silver were removed. Repeating the process a second time, with lines stamped perpendicular to the first set, generated the structure we see here.

Biebuyck, H. A., and G. M. Whitesides, "Self-organization of organic liquids on patterned self-assembled monolayers of alkanethiolates on gold," *Langmuir*, 10 (1994), 1790–93.

Lopez, G. P., et al., "Imaging of features on surfaces by condensation figures," *Science*, 260 (1993), 647–49.

Multiple Experiments

(page 55)

When Xiau Dong Xiang was a research fellow at the Lawrence Berkeley Laboratories, University of California at Berkeley, he supplied me with this sample. The more interesting shot showed some diffraction. The image was taken with daylight coming from a window. In order to get interesting colors, I had to shoot at an angle; I later digitally corrected the distorted perspective so that the rectangles appear straight. Ma, Cl

145

A series of combinations of elements associated with high-temperature superconductivity were sputtered onto a surface in different sequences forming "libraries" of 16 to 128 separate experiments. After deposition, the entire array was heated to 840°C, and then the electrical resistance of the individual spots was measured at temperatures down to 4.2°K to detect electrically superconducting materials. Although no compositions were discovered that were superconducting at temperatures higher than those already known, the best members of the library duplicated existing high-temperature superconducting materials. This strategy is called combinatorial synthesis for its ability to try many possible combinations efficiently and easily, and for the analytical technique used to design the experiments.

Xiang, X.-D., et al., "A combinational approach to materials discovery," *Science*, 268 (1995), 1738–40.

Xiang, X.-D., and I. Takeuchi, *Combinatorial Materials Synthesis*, Dekker, New York (2003).

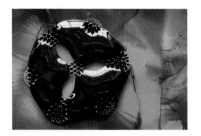

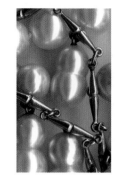

Ferrofluid (pages 56–57)

I dropped a few milliliters of the dark brown liquid on a glass plate and placed that on yellow paper resting on seven circular magnets measuring about 1 cm in diameter each. The magnets created the pattern you see. Ma, Cl

A ferrofluid is a stable, superparamagnetic suspension of a magnetite colloid with sizes of approximately 10 nm—smaller than the domains required for stable ferromagnetism. Ferrofluids are used for a range of applications, including those in which the fluid must be held in a particular location (lubricants for certain types of disk drives and dampers for high-performance speakers).

Raj, K., B. Moskowitz, and S. Tsuda, "New commercial trends of nanostructured ferrofluids," *Indian Journal of Engineering and Materials Sciences*, 11 (2004), 241–52.

Rosensweig, R., *Ferrohydrodynamics*, 2nd ed., Dover Publications, New York (1997).

Small Machines (pages 58–59)

I photographed the blades of a microrotor fabricated by Chunang-Chia Lin from the labs of Professors Alan Epstein, Stephen Senturia, and Martin Schmidt at MIT. Choosing which plane to keep in focus is always a question in microscopic photography. Mi, N

Epstein, A. H., S. D. Senturia, et al., "Micro-heat engines, gas turbines, and rocket engines—the MIT microengine project," American Institute of Aeronautics and Astronautics paper, 1997.

Gabriel, K. J., "Engineering microscopic machines," *Scientific American*, 273 (1995), 150–53.

Jardin, A. P., et al., eds., *Smart Materials Fabrication and Materials for Microelectromechanical Systems*, Materials Research Society Proceedings, 276 (1992).

Pearls (pages 60–61)

These were my grandmother's "peanut" or "double" pearls along with another string of single pearls (see the single pearl with a knot). I shot with diffused daylight with a 105 mm lens. Ma

Sarikaya, M., and I. Aksay, eds., *Biomimetics: Design and Processing of Materials*, American Institute of Physics, Woodbury, N.Y. (1995), 35–90.

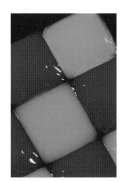

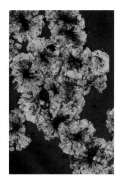

Square Drops of Water

(page 65)

Nicholas Abbott, who was then a postdoctoral fel-low in the Whitesides lab and is now a professor in Chemical and Biological Engineering at the University of Wisconsin-Madison, created my requested patterned self-assembled monolayer, SAM. To emphasize the shape of the drops, we added two fluorescing dyes to the water before I dropped them on the surface. The image appeared on the cover of Science, *September 4, 1992. Ma*

The surface on which the drops rest is a hydrophilic, self-assembled monolayer sup-ported on a thin gold film. Micromachining of 1-μm lines into this surface in a square grid exposed bare gold; these lines of bare gold were covered with a hydrophobic self-assem-bled monolayer. Drops of water containing dyes were then introduced into the square hydrophilic areas. The drops did not mix between squares, despite the small separation between their edges.

Abbott, N. L., J. P. Folkers, and G. M. Whitesides, "Manipulation of the wettability of surfaces on the 0.1–1-micrometer scale through micromachining and molecular self-assembly," *Science*, 257 (1992), 1380–82.

Crystals of Small Particles

(page 66)

I wanted to see how these nanocrystals appeared under an optical microscope. Christopher Murray and Cherie Kagan, then graduate students in Professor Moungi Bawendi's lab, were usually more interested in scanning electron microscopy (SEM) images. Here, I used a blue filter when making the image, which suggested that these crystals show luminescence in the yellow part of the spectrum. Mi

This image is an optical micrograph of ordered crystals of colloidal particles of CdSe, the sur-face stabilized by a thin adsorbed film of an n-alkanethiolate. The colloidal crystals were pro-duced by depositing a drop of a suspension of the colloid in a solution on a fused silica slide and then allowing the solvent to evaporate under reduced pressure.

Murray, C. B., Kagan, C R., and Bawendi, M. G., "Self-Organization of CdSe Nanocrystallites into Three-Dimensional Quantum Dot Superlattices," *Science,* 270 (1995), 1335–37.

Liquid Crystal Film

(pages 68–69)

The sample was mailed from Nicholas Abbott's lab when he was at the University of California at Davis. The grids alone were not enough to make the image interesting, so I composed the image to include the nonpatterned area on the left. Mi, N

Both surfaces of the cell containing the liquid crystal (LC) were self-assembled monolayers (SAMs) supported on semitransparent gold. The pattern in the texture of the liquid crystal was caused by a patterned SAM on one surface of the cell. The disordered regions were caused by competing boundary conditions (the SAM on one surface causing the LC to align parallel to the surface, the SAM on the other surface caus-ing the LC to align normal to the surface). The uniform brown regions have matching surfaces (SAMs on both surfaces cause the LC to align normal to the surface).

147

Brake, J. M., A. D. Mezera, and N. L. Abbott, "Effect of surfactant structure on the orientation of liquid crystals at aqueous-liquid crystal inter-faces," *Langmuir,* 19 (2003), 6436–42.

Collins, P. J., *Liquid Crystals: Nature's Delicate Phase of Matter,* 2nd ed., Princeton University Press, Princeton (2001).

Drawhorn, R. A., and N. L. Abbott, "Anchoring of nematic liquid crystals on self-assembled monolayers formed from alkanethiols on semi-transparent films of gold," *Journal of Physical Chemistry,* 99 (1995), 16511–15.

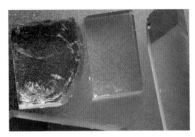

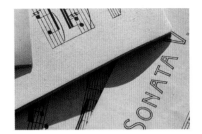

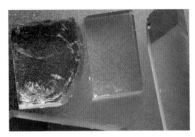

Man-Made Crystal (page 71)

Robert Birgeneau, then Dean of the School of Science at MIT, worked with artificially produced crystals. These were placed on a glass plate, which I thought added nice color to the image. The crystals were photographed under daylight with a 105 mm lens. Ma

These crystals are calcium fluoride and manganese fluoride, materials used in experiments with lasers operating in the ultraviolet (UV). The crystals were grown by top-seeding, in which a small seed crystal is brought into contact with a flux heated to a temperature near the optimal growth temperature. As the heat is lost by radiation or conduction, the flux cools and the seed crystal grows. Heat can be delivered either electrically or by using an infrared laser. High-quality, high-purity crystalline materials are essential to many electronic and optical technologies (for example, silicon-based microelectronics), and the art and science of growing crystals is very highly developed in these areas.

Gilman, J. J., ed., *The Art and Science of Growing Crystals*, John Wiley & Sons, New York (1963).

Yeh, P., *Crystals*, John Wiley & Sons, New York (1988).

Sheet Music (pages 72–73)

I shot this old piece of music outside with a 105 mm lens at around 2:00 in the afternoon. At this time of day, shadows are not so overwhelming that they become more than just an additional compositional element. Ma

Klamkin, M., *Old Sheet Music: A Pictorial History*, Hawthorn Books, New York (1975).

Krummel, D. W., *The Literature of Music Bibliography: An Account of the Writings on the History of Music Printing and Publishing*, Fallen Leaf Press, Berkeley, Calif. (1992).

Player Piano Roll
(pages 74–75)

I unrolled one of the four I bought in an antique store on a glass table. I used a 55 mm lens. Ma, Cl

These rolls were an early form of digital ("hole" vs. "no hole") control of a machine. Many machine controllers in the early days of digital control used tapes containing arrays of holes.

Gill, D., ed., *The Book of the Piano*, Cornell University Press, Ithaca (1981).

Wier, A. E., *The Piano, Its History, Makers, Players, and Music*, Longmans, Green and Co., London (1940).

148

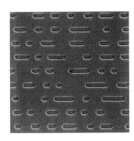

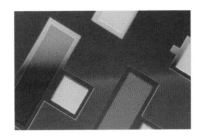

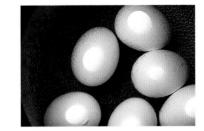

149

Compact Disc (page 76)

Albert Folch, then a postdoctoral fellow in the Department of Electrical Engineering and Computer Science at MIT, helped me create this scanning electron micrograph at the Microsystems Technology Laboratories at MIT. SEM, DC

These disks are another form of digital information storage ("pit" vs. "no pit"). So long as the plastic of which the disc is made is preserved, it is one of the longer-lasting media for information storage.

Baert, L., L. Theunissen, and G. Vergult, eds., *Digital Audio & CD Technology,* 3rd ed., Newnes, Oxford (1995).

May, P. W., "Diamond thin films: A 21st century material," *Philosophical Transactions of the Royal Society of London A,* 358 (2000), 473–95.

Diamond Electron Emitters (pages 78–79)

Michael Geis, then staff scientist at MIT's Lincoln Labs, gave me one of his samples. If you look carefully you can see the nonuniformity of the background emphasized with Nomarski. Mi, N, Cl

New methods for preparing diamonds are a relatively recent development. For years, artificial diamonds have been grown under extremes of temperature and pressure. They can now be grown at atmospheric pressure and relatively low temperature from the vapor phase by chemical vapor deposition, which generates molecular fragments containing carbon atoms and hydrogen atoms. The carbon deposits on a surface as a mixture of amorphous carbon and high-carbon polymers, graphite, and small quantities of diamond. The hydrogen atoms react with all of these, reconverting them to volatile carbon. The reaction is slowest with diamond, which remains when the others are burned away. The cold-cathode emitters shown in this image have an array of holes filled with diamond paste and baked under hydrogen at 1080∞C. Electron emission occurred at very low electric field (0–1 V/µm).

Geis, M. W., and J. C. Angus, "Diamond Film Semiconductors," *Scientific American,* 267 (1992), 84–89.

Eggshells, and Breathing (page 80)

Three loves of my life—science, photography, and cooking—are all represented in this image. While standing over the stovetop in my kitchen with my digital camera attached to my tripod, I took a series of images of eggs boiling. Along with watching the bubbles nucleate, I was also fascinated by the similarity in form between the bubbles and the eggs.

Mortola, J. P., "Metabolic response to cooling temperatures in chicken embryos and hatchlings after cold incubation," *Comparative Biochemistry and Physiology A—Molecular & Integrative Physiology,* 145 (2006), 441–48.

http://exploratorium.edu/cooking/eggs/explore-text.html.

http://newton.ex.ac.uk/teaching/CDHW/egg/.

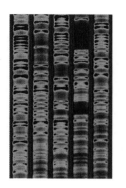

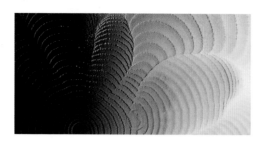

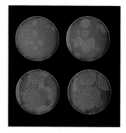

Analysis of DNA (page 83)

I cut a 4 × 5 piece from Giang Zhou's original X-ray film from Professor Philip Sharp's lab in the Department of Biology at MIT. The film recorded DNA sequences. I scanned it on a flatbed scanner and colored the image in Photoshop. DC

Tat Stimulatory Factor 1 (Tat-SF1) is a protein produced by a human cDNA clone that encodes this cellular cofactor for human immunodeficiency virus-1 (HIV-1 Tat). The Tat protein stimulates transcription from the HIV-1 long terminal repeat with the help of several cellular proteins, including Tat-SF1. Tat is absolutely essential for replication of HIV-1. This DNA sequencing gel shows the DNA sequence of the NH2-terminal portion of the Tat-SF1 gene. Many technologies have now been harnessed to sequence DNA in different applications.

Gonick, L., and M. Wheelis, *The Cartoon Guide to Genetics*, Harper Collins, New York (1991).

Sharp, P. A., and R. A. Marciniak, "HIV TAR: An RNA enhancer?" *Cell*, 59 (1989), 229–30.

Venter, J. C., et al., "The sequence of the human genome," *Science*, 291 (2001), 1304–51.

Watson, J. D., et al., *Recombinant DNA*, 2nd ed., Scientific American Books, W. H. Freeman and Co., New York (1992).

Migrating Bacteria

(pages 86–87)

The light for this shot was unusual: to read the contours I had to combine front lighting with irregular back lighting. Using a magnifying mirror that reflected an irregular background, I placed the Petri dish on an angle. Professor James Shapiro, then in the Department of Biochemistry and Molecular Biology at the University of Chicago, supplied me with the samples. Ma, Cl

Bacteria are favored organisms for the study of how a stimulus sensed by an organism and converted into a signal inside the cell results in an observable change in behavior. The responses of bacteria to simple chemical signals are reasonably well understood. The behavior of collections of bacteria is much more complicated. The beautiful regularity and complexity of this pattern of *Proteus* colonies was unexpected. One key to the formation of the visible terraces is the density of growth: when the population of bacteria reaches a certain level, it differentiates a limited subpopulation of swarmer cells that move out and colonize a new region.

Shapiro, J. A., "Bacteria as multicellular organisms," *Scientific American*, 256 (1988), 82–89.

Shapiro, J. A., "Bacteria are small but not stupid: Cognition, natural genetic engineering, and sociobacteriology," http://shapiro.bsd.uchicago.edu/2006.ExeterMeeting.pdf.

Shapiro, J. A., and M. Dworkin, eds., *Bacteria as Multicellular Organisms*, Oxford University Press, New York (1997).

Oscillating Chemical Reaction (pages 88–89)

Matthew Eager, then an undergraduate student, helped set up the reaction in Professor Anatole Zhabotinsky's lab in the Department of Chemistry at Brandeis University. We placed the Petri dish containing the reactants on a light table and surrounded it with black cloth. I took an exposure every 11 seconds and was amazed at the regularity of the oscillations. These eight images are an edited representation of the series I shot.

These images are of the Belousov-Zhabotinsky (BZ) reaction, an example of a so-called "oscillating chemical reaction." The reactants are malonic acid, and bromate ion in acid solution. The colored bands are due to an indicator dye that indicates the oxidation potential of the solution. The overall process leading to the oscillating bands requires at least four coupled chemical reactions. It is possible to simulate the basic patterns generated in these systems by computer. An accurate description of the reaction requires integrating a set of twenty-five or thirty reactions. The dots in the image are an artifact: they are bubbles of gas that have formed in the reaction medium.

Epstein, I. R., and J. A. Pojman, *An Introduction to Non-Linear Chemical Dynamics*, Oxford University Press, New York (1998).

Vanag, V. K., L. F. Yang, M. Dolnik, A. M. Zhabotinsky, and I. R. Epstein, "Oscillatory cluster patterns in a homogeneous chemical system with global feedback," *Nature*, 406 (2000), 389–91.

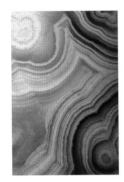

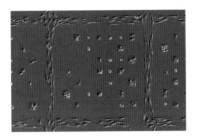

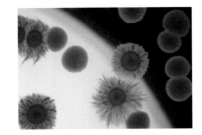

Agate (page 91)

Using auxiliary tungsten lighting for the stereomicroscope, I photographed a sample of agate from my colleague Elizabeth Connors' mineral collection. SM

Heaney, P. J., and A. M. Davis, "Observation and origin of self-organized textures in agates," *Science*, 269 (1995), 1562–65.

Mammalian Cells on a Patterned Surface (pages 92–93)

Christopher Chen, then a graduate student working with Professor Donald Ingber at Children's Hospital in Boston, gave me these samples. With Nomarski, the cells' outlines appear more pronounced than in other images I made. Mi, N

The surface was patterned by microcontact printing two types of self-assembled monolayers on gold. The surfaces to which the cells were to attach were covered with a monolayer of organic molecules that terminated in methyl ($-CH_3$) groups. This surface adsorbed proteins that permit cellular attachment—fibronectin, laminin. The remaining surface was covered with another molecule terminating in oligo(ethylene glycol) groups ($-CH_2CH_2O)_nOH$). Neither proteins nor cells attach to these surfaces. The cells were bovine capillary endothelial cells. They were plated in serum-free media, cultured overnight, and fixed in 4 percent formaldehyde in phosphate-buffered saline solution. This technique is now widely used in mammalian cell culture to examine the movement and internal structure of cells.

Chen, C. S., M. Mrksich, S. Huang, G. M. Whitesides, and D. E. Ingber, "Geometric control of cell life and death," *Science*, 276 (1997), 1425–28.

Weibel, D. B., and G. M. Whitesides, "Applications of microfluidics in chemical biology," *Current Opinion in Chemical Biology*, 10 (2006), 584–91.

Colonies of Yeast (page 94)

I asked Julie Kohler, then a research fellow in Professor Gerald Fink's lab at the Whitehead Institute, to grow both types of colonies on one Petri dish. Ordinarily, the procedure was to grow them separately. This particular stereo microscope gave me more control over the light source than newer versions. I prefer this older type of microscope because of that ability. The light in this case was reflected off a mirror. SM

Wild-type *Candida albicans* was modified by homozygous disruption of the CPH1 gene. This gene is homologous to a gene (Ste12p) in *Saccharomyces cerevisiae* (baker's yeast) known to be involved in the yeast pheromone response pathway and important in mating of this yeast. This image is of wild-type and genetically modified strains of *C. albicans* using conditions that cause expression of the filamentous morphology; only the wild-type yeast is able to respond to these conditions. These experiments help to understand how cells sense their environment, wage war with one another, and locate areas hospitable to growth.

Liu, H., J. Kohler, and G. R. Fink, "Suppression of hyphal formation in *Candida albicans* by mutation of a STE 12 homolog," *Science*, 266 (1994), 1723–26.

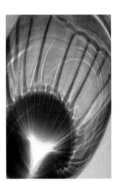

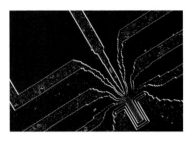

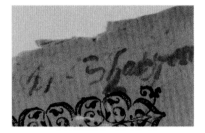

Tears of Wine (page 97)

After a series of attempts at directly photographing the glass of wine, I discovered that the shadowgram gave the best portrayal of the Marangoni effect.

The origin of surface tension is the tendency of molecules to surround themselves with other molecules. Molecules at a liquid–vapor or solid–vapor interface do not enjoy the energetic stabilization that would come from being completely surrounded by others. The Marangoni effect is the effect responsible for "tears" or "legs" of wine, and is one of the most enjoyable of the manifestations of surface tension. The surface tension of a solution of ethanol in water is substantially lower than that of pure water. The water-ethanol mixture coats the surface of the wine glass when it is swirled, and the large surface area of the thin liquid film allows rapid volatilization of the ethanol from the uppermost portions of the film and an accompanying rapid increase in the surface tension. This increase in surface tension is reflected in a tendency for the liquid to pull itself into drops, to minimize the surface area of the low-ethanol content surface film of the liquid.

Adamson, A. W., and A. P. Gast, *Physical Chemistry of Surfaces*, 6th ed., John Wiley & Sons, New York (1997).

Microelectrodes (pages 98–99)

A similar shot appeared on the cover of Langmuir, *October 1995. The work was completed in the laboratory of Professor Mark Wrighton, former provost of MIT, and now chancellor of Washington University, St. Louis. Mi, N*

Tatistcheff, H. B., I. Frisch-Faules, and M. S. Wrighton, "Comparison of diffusion constants of electroactive species in aqueous fluid electrolytes and polyacrylate gels: Step generation-collection diffusion measurements and operation of electrochemical devices," *Journal of Physical Chemistry,* 97 (1993), 2732–39.

Shakespeare's Signature (page 101)

Frank Mowery, head of conservation at the Folger Shakespeare Library in Washington, D.C., introduced me to the library's glorious Shakespeare collection. This signature is one of very few in the world. I used the 105 mm lens on a copy stand. Ma

The book from which this signature was taken is *Archaionomia* ("ancient laws"), printed in London in 1568 and found in the Folger Shakespeare Library in Washington, D.C. It is a text of old English laws accompanied by Latin translations by William Lambard. Mowery believes that the signature is authentic because it was found before the earliest of the known signatures, resembles it, and is known to have been written long before the time of the notorious Shakespeare forger William Henry Ireland. Mowery believes that, if genuine, this signature may suggest that Shakespeare had studied as a law clerk under Lambard. Shakespeare might have been given this book by the author.

Langwell, W. H., *The Conservation of Books and Documents*, Greenwood Press, Westport (1974).

Plenderleith, H. J., *The Conservation of Antiquities and Works of Art: Treatment, Repair, and Restoration*, 2nd ed., Oxford University Press, London (1971).

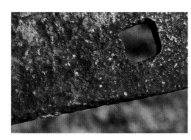

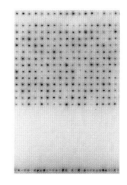

Rust (page 104)

This is a detail of a wonderful old farm implement. I shot the square hole with a 105 mm lens in broad daylight.

Atkins, P. W., *The Second Law*, Scientific American Books, W.H. Freeman and Co., New York (1984).

Fenn, J. B., *Engines, Energy, and Entropy*, W.H. Freeman and Co., New York (1982).

Von Baeyer, H. C., *Maxwell's Demon: Why Warmth Disperses and Time Passes*, Random House, New York (1998), 165.

Smith, E. B., *Basic Chemical Thermodynamics*, 5th ed., Imperial College Press, London (2004).

Mechanical Failure of a Thin Film (pages 106–107)

Errors can be interesting and this one produced a remarkable consistency in the size of the peeled film cylinders. This observation led me to ask more questions about fracturing. Mi, N

Ehrlich, D. J., and J. Melngailis, "Fast room-temperature growth of SiO_2 films by molecular layer dosing," *Applied Physics Letters*, 58 (1991), 2675.

Keckes, J., "Simultaneous determination of experimental elastic and thermal strains in thin films," *Journal of Applied Crystallography*, 38 (2005), 311–18.

Microexplosions Inside Silica (page 109)

I first imaged the quartz sample with transmitted light but later found that epi-illumination gave a better picture. I color-inverted the whole image in Photoshop so the structures (the holes) would be easier to see. If the purpose of the image is to show the formation of the cavities, then the color inversion is just, as long as I make that information clear to the viewer. Mi, Cl

These cavities were made by focusing the light from a pulsed laser on the interior of a transparent silica slab. Nonlinear absorption of light created a small region of plasma—a gas of electrons and ions—in the silica, and left a small hole in the silica. The optical microscope images the damage—small fractures, local changes in index of refraction—left by this event.

Glezer, E. N., et al., "3-D optical storage inside transparent materials," *Optics Letters*, 21 (1996), 2023–25.

153

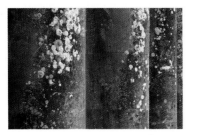

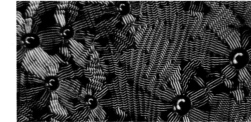

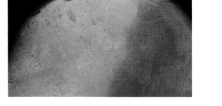

Columns with Lichen

(pages 110–111)

Lichens grew on just one side of the columns on one of the structures at Stourhead. Professor Donald Pfister, director of the Harvard University Herbaria, confirmed that I had, in fact, photographed lichen.

The lichens attacking these columns are several species of crustose lichens. Crustose is a type of lichen thallus that attaches firmly to the surface of the substrate.

Barinaga, M., "Origins of lichen fungi explored," *Science,* 268 (1995), 1437.

Lutzoni, F., M. Pagel, and V. Reeb, "Major fungal lineages are derived from lichen symbiotic ancestors," *Nature,* 411 (2001), 937–40.

Wrinkled Gold (pages 112–113)

Looking at the sample under the microscope with reflected light, I became fascinated by the differences between the herringbone and radial patterned buckling. Mi, N

A slab of polydimethylsiloxane (PDMS) that had a flat top was placed in an electron-beam evaporator configured to evaporate gold. Heat radiated by the very hot gold caused the surface of the PDMS to warm, and expand. Deposition of gold on the PDMS thus occurred on a thermally expanded surface. When the deposition was completed and the e-beam extinguished, the PDMS cooled and contracted. Since the coefficient of expansion of PDMS is much larger than that of gold, the gold film was put under compression. The stress of this compression was relieved by buckling. The large gold balls (inadvertently ejected from the surface of the heated gold by too-rapid heating) provided nucleation points for this buckling.

Bowden, N., S. Brittain, A. G. Evans, J. W. Hutchinson, and G. M. Whitesides, "Spontaneous formation of ordered structures in thin films of metals supported on an elastomeric polymer," *Nature,* 393 (1998), 146–49.

Oxide Layer on a Copper Pan (pages 114–115)

I borrowed this magnificent au gratin dish from a friend and shot its oxidized underside using daylight with my 105 mm lens.

Fromhold, Jr., A. T., *Theory of Metal Oxidation,* North Holland Publishing Co., New York (1975).

Over, H., and A. P. Seitsonen, "Oxidation of metal surfaces," *Science,* 297 (2002), 2003.

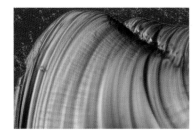
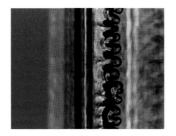
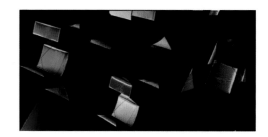

Surface of Broken Glass

(page 116)

I was inspired by one of my undergraduate photography students at MIT, Katherine Notter, who took a wonderful picture of a piece of broken glass, so I thought I'd give it a try. Mi, N

Fracture mechanics is the study of the mechanisms of failure of solids. The wide range of failure mechanisms includes not only cracking but also solids, flowing, slipping along planes in crystal; or generating bubbles by cavitation ahead of the crack tip. Glass is a brittle solid and fails by concoidal fracture in a series of progressive fractures of the silicon-oxygen and boron-oxygen bonds that make up the glass, interspersed with pauses while the tension builds to levels that permit the fracture tip to advance. This picture was taken using reflection Nomarski-mode photomicrography, and the colors provide a map of the topography of the surface. Reflection Nomarski mode records lateral gradients in the phase, rather than the amplitude, of the light; these phase shifts are translated into differences in color and prove a map of the surface coded in terms of height. These techniques of microscopy are also widely used in biology to provide detail about transparent samples (cells and tissues).

Abraham, F. F., "Unstable crack motion is predictable," *Journal of the Mechanics and Physics of Solids*, 53 (2005), 1071–78.

Marder, M., "Energetic developments in fracture," *Nature*, 381 (1996), 275–76.

Failure in Adhesion

(pages 118–119)

Professor Manoj Chaudhury and Bi-Min Zhang Newby, then a graduate student, at the Department of Chemical Engineering at Lehigh University sent me some samples of surfaces. As they suggested, I stuck some transparent tape on one, slowly pulled it away, and observed the adhesion changes at the interface of the adhesive and the surface under the microscope as it was being pulled. Mi, N

A pressure-sensitive adhesive like 3M's Scotch™ tape is a viscoelastic polymer. Any elastic deformation within the adhesive is resisted by its high viscosity. When the adhesive is peeled from a solid surface, competition between applied pressure and surface tension in the viscous polymer leads to an instability of the meniscus that grows into waves and well-developed fingering patterns whose shape and size depend upon the interaction of the adhesive with the substrate.

Chung, J. Y., and M. K. Chaudhury, "Roles of discontinuities in bio-inspired adhesive pads," *Journal of the Royal Society Interface*, 2 (2005), 55–61.

Dillard, D. A., A. V. Pocious, and M. Chaudhury, eds., *The Mechanics of Adhesion*, Elsevier Science, Amsterdam (2001), vol. 1.

Newby, B. Z., M. K. Chaudhury, and H. R. Brown, "Macroscopic evidence of the effect of interfacial slippage on adhesion," *Science*, 269 (1995), 1407–09.

Prismatic Soap Bubbles

(pages 122–123)

James Ossi called himself a gismologist. He specialized in creating bubble machines. This one used to sit in the physics department lobby at MIT. I was always mesmerized when I passed by it. At that time, still using film, I "pushed" Velvia film a full stop for this three-image composite. The bubbles were continuously moving and I needed the faster shutter speed. I shot with a 55 mm lens through the glass. Ma

Bubbles have been a laboratory for surface science since its beginning.

Isenberg, C., *The Science of Soap Films and Soap Bubbles*, new ed., Dover Publications, New York (1992).

Lovett, D., *Demonstrating Science with Soap Films*, Institute of Physics Publishing, Bristol, U.K. (1994).

155

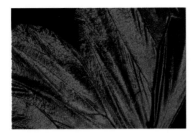 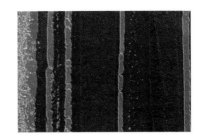 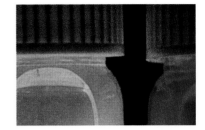

Wing of a Morpho Butterfly (page 125)

I bought this Morpho sulkowski from Brazil in a store in San Francisco and photographed it with simple daylight from a window. Ma

The blue colors of the wing of the morpho butterfly are the result of interference in an elaborately structured set of scales on the wing. The blue is entirely reflected light. The color of the wing (in the absence of the reflections) is a dull grey-brown.

Silver, J., R. Withnall, T. G. Ireland, et al. "Novel nano-structured phosphor materials cast from natural Morpho butterfly scales," *Journal of Modern Optics*, 52 (2005), 999–1007.

http://webexhibits.org/causesofcolor/15.html.

Peterson, I., "Butterfly blue: Packaging a butterfly's iridescent sheen," *Science News*, 148 (1995), 296–97.

Colors from Transparent Spheres (pages 126–127)

Professor Kuniaki Nagayama at the University of Tokyo in the Department of Life Sciences sent me various samples from his lab in Japan. The green image on p. 127, taken with a stereo microscope, was shot with reflected tungsten light. The larger brown and purple microscopic image was taken with reflected light using Nomarski. Mi (green image) SM, Mi, N (brown and purple image)

The preparation of uniformly sized polymer is usually carried out by emulsion polymerization. The polymerization is initiated simultaneously thermally or photochemically, and the rate of growth of polymer becomes mass-transport-limited under appropriate conditions: under these circumstances, the length of the polymer chain and the mass of polymer are independent of the size of the original droplet. Forming ordered monolayers of these colloids is an art. Here, the microspheres deposited on a surface from a liquid that was being withdrawn; the gentle wiping action of the retreating liquid drop compacted the microspheres and allowed them to crystallize.

Kralchevsky, P. A., and K. Nagayama, "Capillary interactions between particles bound to interfaces, liquid films and biomembranes," *Advances in Colloid and Interface Science*, 85 (2000), 2–3, 145–92.

Dots on Paper (pages 128–129)

I photographed a published image of the suspensions of fluorescent particles (see page 22) from a Technology Review article, May/June 1996. SM

Eckstein, H. W., *Color in the 21st Century: A Practical Guide for Graphic Designers, Photographers, Printers, Separators, and Anyone Involved in Color Printing*, Watson-Guptill Publications, New York (1991).

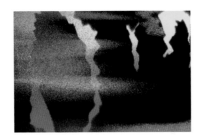

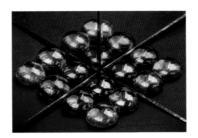

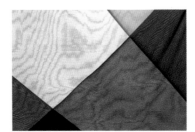

Hologram (pages 130–131)

Betsy Connors, an artist in Somerville, Massachusetts, used holography to create an environmental piece called Light Forest, a Holographic Rainforest. *It is impossible to capture the three-dimensional quality of a hologram in two dimensions, so I focused in on the plastic surface detail. The granulated effect is the scattering of the light. Ma*

A hologram replicates the pattern of light in space that would be generated by light scattered by a three-dimensional object. Most techniques for storing information holographically rely on interference between a reference source of light and light from the storage medium that contains the information. This medium is an analog of a diffraction grating, storing information about the amplitude and phase of the light relative to the reference beam. This information can be recorded as a reflective surface on a sheet of plastic, or as variations in the index of refraction in a three-dimensional solid.

Calfield, H. J., "Holograms," *National Geographic*, 165 (1984), 364–77.

Kasper, J. K., and S. A. Feller, *The Complete Book of Holograms*, John Wiley & Sons, New York (1987).

Saxby, G., *Practical Holography*, 3rd ed., IOP Publishing, Washington, D.C. (2003).

Reflections (page 133)

Four round colored pieces of glass from a dear friend and two mirrors made a challenging photographic exercise. The set-up was under the skylight in my studio, and I used my 105 mm lens on a digital camera.

Understanding reflections has been a puzzle since the discovery of mirrors, and discussions of the apparent transposition of right and left, but not of top and bottom, continue to this day.

Gregory, R., *Eye and Brain: The Psychology of Seeing*, 5th ed., Princeton University Press, Princeton, N.J. (1998).

http://www.glenbrook.k12.il.us/gbssci/phys/mmedia/optics/lr.html.

http://www.phy.ntnu.edu.tw/oldjava/optics/mirror_e.html.

Moiré Pattern (page 134)

Draping and overlapping a few loosely woven fabrics and attaching them to my window during the day created some interesting moiré patterns.

Kaura, S. K., D. P. Chhachhia, and A. K. Aggarwal, "Interferometric moiré pattern encoded security holograms," *Journal of Optics A—Pure and Applied Optics*, 8 (2006), 67–71.

Kinneging, A. J., "Demonstrating the optical principles of Bragg's law with moiré patterns," *Journal of Chemical Education*, 70 (1993), 451–53.

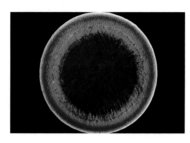

acknowledgments

Light Reflecting from a Polymer-Containing Gel (page 137)

This image is one of a series taken over a period of 24 hours. (On page 6, you'll see the final image.) Over time, the colors shifted. The image was captured with my digital camera with a 105 mm lens.

A sample of a very high molecular-weight block co-polymer (poly(styrene-*b*-isoprene)) was dissolved in cumene and placed in a thin film between a glass slide and a glass cover slip. The colorless polymer phase-separated into a layered structure, with layers parallel to the plane of the slide. The separation between layers was in the order of 200–300 nm. These layers acted as a multilayer dielectric mirror—a little like the oil slick discussed earlier. The details of the processes that give slightly greater layer-to-layer separations at the periphery of the disk than under its center are not clear, but probably have to do with mass transport coupled to evaporation of solvent from exposed solution at the edges of the glass cover slip.

Lee, W., J. Yoon, H. Lee, and E. Thomas, "Direct 3-D imaging of the evolution of block copolymer microstructures using laser scanning confocal microscopy," *Macromolecules*, 40, (2007), 6021–24.

This book describes many individuals' work. Without their help in obtaining the images and understanding the objectives of the research, we could not have proceeded. In the "Notes and Readings" section, we acknowledge the individuals with whom we worked most closely; here, we thank all of them collectively for making the science that we describe happen: this book celebrates their accomplishments. We note the obvious—that the understanding of nature grows through many contributions over many years, and in praising explicitly a few of those now involved in the process, we also mean to salute their colleagues, coworkers, competitors, and forebears.

The John Simon Guggenheim Memorial Foundation supported the photography ten years ago. Without this support, there would have been no book.

Finished prose is a collaboration between writer and editor. We were most fortunate in this work in having two editors for the first printing—Jisho Cary Warner and Barbara Whitesides—each with an affinity for clarity, an ear for tone, and an instinct for science; Barbara Whitesides has edited the text that accompanies the images new to this edition.

We thank Kate Brick at Harvard University Press for reassembling the disassembled pieces of the revised book into a coherent whole. The book would be different and much the poorer had they all not taken it in hand. Others have also helped. Among them we especially thank Mary Cattani, Ken, Matthew, and Michael Frankel, Greer Gilman, Giselle Weiss, and Ben and George Whitesides. Stuart McKee and Jill Jacobson initially integrated the images and the text and Tim Jones did the same for this edition. Part of the book was written while George Whitesides was at Cambridge University as the Lord Todd visiting professor of chemistry.

Our colleagues at MIT and Harvard built the intellectual community in which we lived while working on this project. They have taught us most of what we know of science and technology.